# THE POLITICIAN'S GUIDE

## TO ASSISTED SUICIDE, CLONING,

## AND OTHER CURRENT CONTROVERSIES

# THE POLITICIAN'S GUIDE

## TO ASSISTED SUICIDE, CLONING,
## AND OTHER CURRENT CONTROVERSIES

George J. Marlin

*Foreword by*
Jeremiah A. Denton Jr.
Rear Admiral, U.S. Navy

MORLEY BOOKS ❖ WASHINGTON, D.C.
1998

*Library of Congress Cataloging-in-Publication Data*

ISBN 0-9660597-1-9

Printed in the United States of America

My efforts are dedicated to

Mike Long,

*street-corner conservative*

and

Patrick Foye, Esq.,

*defender of subsidiarity*

# CONTENTS

vii

PART III

EUGENICS AND CLONING

PART IV

THE DEATH PENALTY

# FOREWORD

A FEW YEARS BACK, during an extended stay as a guest of the communist government in Hanoi, I had time to think about the differences between a free society, such as America has always had, and a totalitarian society, such as Vietnam still has. Seven years and seven months as a prisoner of war allow a man plenty of time to think. Under the circumstances (which, courtesy of my hosts, included extended periods of solitary confinement, frequent episodes of torture, and an almost constant state of starvation), I can probably be forgiven if the America I imagined was just a little idealized.

And yet this does not explain the culture shock I experienced upon returning home. When I left the United States in 1965, ours was a nation at the beginning of a long process of self-examination, which was a good thing. But the America I found in 1973 was no longer searching its soul, it was gazing into its navel, which is a bad thing. Self-indulgence is poison to a democracy.

As George Marlin makes clear in this book, our Founding Fathers based their whole brave experiment upon the premise that American citizens fervently believe in God. To be an American is to believe that there is a power greater than any man or any law. Americans have always understood that virtuous men and good laws are such because their lives and principles serve a higher truth.

What struck me in 1973 and continues to trouble me twenty-five years later is the ease with which self-indulgence discards our most honored traditions and the norms that go with them. This can only be because New-Age denizens no longer believe we are "one nation under God." We used to believe in the sacredness of innocent life, yet now many willingly kill the unborn, the unwell, and the unyoung. We used to believe that the guilty should be punished, but now thieves are set free to steal again, rapists to rape again, and murderers to kill again. And whereas we once thought every man and woman uniquely fashioned in God's image, now we seem hell-bent upon using genetic science as a kind of horrendous cookie cutter to remake Man in our own likeness.

As I said: self-indulgence is poison in a democracy. But I do not despair. If, like me, you are a religious person you can derive confidence about the future from Christ's promise about the Church: that even the gates of hell will not prevail against it. This is a great comfort, although we must note that His words were spoken about the Church and not about America. When we say "God bless America," as I did when I stepped off the plane that brought me out of captivity in Vietnam, we speak a prayer that so far has been answered. And yet we must remember that  our glorious past is not absolute insurance against an accursed future. I always shiver when I recall President Lincoln's reply when he was asked about the cause of the Civil War. "We have forgotten God," he said.

What you are holding in your hands is an antidote to our cultural malaise. At the heart of George Marlin's view is a principle that every law needs to reflect and every law-maker (lawyer, judge, or legislator) must reflect upon: law is a part of morality. It ought never to be legally right to do what is morally wrong. Only a healthy respect for natural law can keep in balance the conflicting claims of the State and the Self.

This may be a bitter pill for some, but we all need to swallow it.

JEREMIAH A. DENTON JR.
REAR ADMIRAL, U.S. NAVY

*Admiral Denton, former United States Senator from Alabama, is author of the acclaimed memoir* When Hell Was in Session.

# PREFACE

*We have now sunk to a depth at which re-statement of the obvious is the first duty of intelligent men.*

George Orwell

I CONSIDER MYSELF a public servant more than a politician. I have had the chance to run for mayor of New York (as the Conservative Party's candidate in a three-way race against David Dinkins and Rudolph Giuliani), and I have had the privilege to serve as Executive Director of one of the nation's oldest public agencies, The Port Authority of New York and New Jersey. Both as a candidate and as an administrator I advocated the same fundamental principle: government must be more efficient, less intrusive, and more accountable to the people it serves.

Although I was able to achieve considerable success in reorganizing the Port Authority, I was also unable to completely overcome what economist Milton Friedman has called the "tyranny of the status quo." It will probably come as no surprise to most readers to learn that very few bureaucrats care much about philosophical concepts or that most of them care only about preserving their power, perks and pensions. When I set about to privatize some of the Port Authority's functions, to downsize the agency's massive staff, and to emphasize the importance of individual responsibility and accountability, I was met with raging hostility (occasionally open but more often concealed), and it became clear early on that reform was anathema. (I have always admired G. K. Chesterton's definition of reform as "a metaphor for reasonable and determined men: it means that we see a certain thing out of shape and we mean to put it into shape. And we know what shape.") The truth is that many, if not most, bureaucrats actually prefer the murky mazes of big government, because the labyrinths of rules and regulations conceal their waste and inefficiency and shield them from the scrutiny of "meddling interlopers," which is how they refer to the people's elected officials and their appointed representatives.

While I was prepared for the bureaucrats' "insolence of office" (as Shakespeare called it), I will confess that I was taken aback to find that so many of those elected officials and their appointed representatives were ignorant of the historical and philosophical background of the life and death issues they are called upon to debate and decide. And this included a number of prominent spokesmen who argue for or against assisted suicide, euthanasia, cloning, and the death penalty. It is disturbing to discover how often these eloquent masters of the sound bite are unaware that ideas such as liberty, justice, punishment, human nature, and the common good have both antecedents and consequences. It does not take long in a conversation with some politicians to realize that this depth of understanding is missing, and I was often reminded of the observation that you cannot very well reason with a man about his opinions if he did not use reason to arrive at them.

This is why I have written this book: to provide intelligent politicians (and others) with philosophical and historical perspectives on some of the toughest issues we Americans face. Although my own opinions are fairly obvious, I have attempted to cite authorities on both sides of every controversy, and the reader remains free to decide for himself. It cannot be otherwise.

But it will become evident to all, I think, that every position on each of the "life" issues discussed herein has a common denominator—a concept of the person. For those who believe that man is merely a machine or an animal, it will be easy to rationalize suicide or cloning. For those who believe that man is special—is a person with a soul and, therefore, unique among God's creations—it will be difficult to form a rational justification of suicide or cloning.

As we look back at the twentieth century, we may conclude that the more horrific excesses of recent history have arisen in the name of the kind of ideological fanaticism that routinely discards the sanctity of the human person. I am tempted to thank God that this benighted century is coming to a close, until I recall that—even with the defeat of Nazism and the collapse of socialism—human nature remains unchanged and prospects for the furtherance of human dignity continue to be chancy at best. It is easy to see how modern gnostics such as Hitler, Stalin, Mao, and Pol Pot were so willing to destroy individual human beings in pursuit of their totalitarian ends, but we may find it hard to accept that the very same ends may be achieved by democratic means. But without respect for personhood, for the certitude that every man and woman matters, liberty will become license, and the responsibility to do what is right will degrade into the right to do what is irresponsible. Transcendent values will be replaced by fleeting tastes; the common good will be displaced by raw power. Whether that power turns out to be individual or collective will hardly matter.

Throughout most of this century the American debate about human nature has been dominated by the utilitarian or positivist schools of thought. The origins of these philosophies go back more than a thousand years, but their arguments did not begin to become persuasive to the American mind until after Oliver Wendell Holmes began his thirty-year tenure on the Supreme Court in 1902. What follows is partly a refutation of Holmes' pontifications, but it is mostly a defense of the natural-law tradition. That tradition forms the basis of American democracy, and both assumes and assures the dignity of the human person. As I hope to show, natural law is often our only defense against injustice and inequity.

I AM GRATEFUL to those who assisted and guided me in the preparation of this book. Very special thanks must go to my wife, Barbara Marlin, whose contributions to the assembling of the manuscript were invaluable. For their critiques of the manuscript at various stages of its completion, I want to thank the following: Brad Miner, Richard Rabatin, Pat Foye, Joe Mysak, Charles Woram, James Schall, S.J., Robert Royal, Christine Dumas, M.D., Heather Higgins, and Jack Swan.

Heartfelt thanks as well go to Paul Atanasio, Allen Roth, Dominick Antonelli, Ray O'Sullivan, Gary Eggers, Bob Luckey, Gene McCue, Tom Houston, Deal Hudson, Casey Carter, Eden Simmons, Mike Crofton, and Jeff McGill.

Finally, I offer my profound gratitude to Larry Azar, emeritus professor of philosophy at Iona College. It was Dr. Azar who introduced me to the concepts of natural law, the common good, and subsidiarity.

I acknowledge my indebtedness to each of these people. I hope their kindness is vindicated. I alone am responsible for any errors, inaccuracies, or follies in what follows.

GEORGE J. MARLIN
THE CITY OF NEW YORK
DECEMBER, 1997

# PART I

# NATURAL LAW

# 1

# THE AMERICAN CREDO

*A just law is a man made code that squares with the moral law or the law of God. . . . An unjust law is a code that is out of harmony with the natural law.*

Martin Luther King, Jr.

FROM THE VERY BIRTH of our republic, the American credo has been rooted in the tradition of natural law; has been imbued with the belief that there is a higher standard by which all man-made rules must be measured. In the original draft of the Declaration of Independence, Thomas Jefferson (1743-1826) justified the American case for separation from Great Britain with a classic appeal to the natural law:

> When in the course of human events it becomes necessary for a people to advance from that subordination in which they have hitherto remained & to assume among the powers

of the earth the equal & independent station to which the *laws of nature & of nature's god* entitle them, a decent respect to the opinions of mankind requires that they should declare the causes which impel them to change.

We hold these truths to be *sacred & undeniable*; that all men are created equal & independent, that from the equal creation they derive *rights inherent and inalienable*, among which are the preservation of life & liberty, & the pursuit of happiness . . .[1] [Italics added]

In Jefferson's view, the American Revolution did not break lawful ties to a sovereign realm, but reclaimed transcendent liberties from an illegitimate and corrupt monarchy.[2]

Just six years later in his *Notes on the State of Virginia* Jefferson wondered what *disbelief* in natural law might mean for America's future. "Can the liberties of a nation be secure," he asked, "when we have removed a conviction that these liberties are the gift of God?"[3]

From Thomas Jefferson to Martin Luther King, Jr., Americans have understood that these "laws of nature and of nature's god" are the great guardians of the soul of democracy, which is the intrinsic value of the person. Yet for every proponent of the higher law there is a lawyer or politician or academic who subscribes to one of the ideologies, such as positivism or utilitarianism, that vigorously reject the natural law, and it is not uncommon to read today that belief in natural law is all but dead. But as philosopher Etienne Gilson once quipped, "the natural law always buries its undertakers." How could it be otherwise, given that this law is, in the words of Saint Augustine, "written in the hearts of men, which iniquity itself effaces not."

Indeed there are times when the natural law is our *only* defense against iniquity. This was made evident during the prosecution of Nazi criminals at the 1946 War Crimes Tribunal in Nuremberg, Germany.

## 2

# NATURAL LAW ON TRIAL:
# NUREMBERG AND THE HIGHER TRUTH

*All attempts at passive and active resistance to the
[Nazi] regime were necessarily grounded on natural
law ideas or on divine law, for legal positivism as
such could offer no foundation.*

Heinrich Rommen

AMERICA HAD BEEN at war only a few months when Franklin D. Roosevelt grasped the magnitude of the atrocities being committed in Germany. The president made numerous appeals to the Nazis to spare the lives of innocent people, but by October of 1942 the pleas had given way to a frank warning: "It is our intention," the president stated, "that just and sure punishment shall be meted out to the ringleaders responsible for the organized murders of thousands of innocent persons in the commission of atrocities which have violated every tenet of the Christian faith." [1]

Winston Churchill agreed. Writing to Foreign Minister Anthony Eden in July of 1944, the Prime Minister declared: "There is no doubt that this is probably the greatest and most horrible crime ever committed in the whole history of the world, and it has been done by scientific machinery by nominally civilized men in the name of a great state. . . . It is quite clear that all concerned who may fall into our hands, including the people who only obeyed orders by carrying out the butcheries, should be put to death after their association with the murders has been proved." [2]

In fact, Churchill took the position that the Nazi leaders who were caught should be summarily charged and immediately executed. [3] But others believed strongly that the war criminals ought to stand trial. "Not to try these beasts," wrote American Lieutenant Colonel Murray C. Bernays, who was one of those assigned to collect evidence of Nazi crimes, "would be to miss the educational and therapeutic opportunity of our generation. . . . They must be tried not alone for their specific aims, but for the bestiality from which these crimes sprang." [4]

After much bickering among the Allied powers the War Crimes Tribunal was established, and twenty-one Nazi leaders were charged according to a basic three-part indictment. Part one accused the defendants of conspiracy to wage a war of aggression in violation of international treaties; these were "crimes against peace." Part two charged the defendants with violations of the laws and customs of war as embodied in the Hague and Geneva Conventions and as recognized by the military forces of all civilized nations; these were "war crimes." And, finally, part three accused the Nazis of the extermination of racial, ethnic, and religious groups and with other atrocities against civilians; these were "crimes against humanity." [5]

The arguments of the defense were essentially two. First it was claimed that at the time the various acts were alleged to have been committed, the "crimes" with which the defendants were now being charged had no statutory basis, either in German law or in international law; that the legal basis of the indictments had been created after the fact. Since "ex post facto" laws are constitutionally prohibited by each of the Allied powers, they could hardly have validity in a court convened by those nations. This view had some surprising supporters around the world, ranging from French leftist philosopher Jean-Paul Sartre to American conservative politician Robert Taft, neither of whom had the slightest sympathy for the Nazis.

The defense's second argument was the assertion that the accused ought not to be charged with the consequences of following the orders of Germany's lawful leaders. One defendant, Nazi jurist Hans Frank, had put forward the substance of this defense a decade earlier. In 1935 he had written: "Formerly we were in the habit of saying: 'This is right or wrong.' Today we ask the question: 'What would the Führer say?'" Frank went so far as to assert that love of the leader "has become a concept of law," and his was not an isolated view. [6] In 1936 the Third Reich's chief legal officer decreed that the Führer's decisions were beyond the scrutiny of judges. This was the culmination of the *Führer prinzip*, the leader principle, and it literally allowed Adolph Hitler to become a law unto himself. [7] This being the case at the time, the Nuremberg defense asked, how could obedience to the law—to Hitler and Nazism—be considered illegal? Hitler's authority had been confirmed equally by appointed judges and elected legislators, and he was able to boast of his Nazi party: "We stand absolutely as hard as granite on the ground of legality." [8]

It was, however, a perverse legality. Beginning with the Decree for the Protection of the People of State (1933), which obliterated the personal freedoms formerly protected by the Weimar

Constitution, the Nazis promulgated a series of legal outrages. There were a series of "racial-purity" laws that forbade sexual relations (marital and extramarital) between Jews and non-Jews,[9] imposed the obligation of a license certifying that partners were racially "fit to marry,"[10] and that detailed punishments for anyone "who contributes . . . to the racial deterioration and dissolution of the German people by interbreeding with Jews or the colored races."[11] There were laws that forced registration of "alien races" and genetically "less valuable" individuals; laws that sought to strip citizenship from Eastern European Jews who had emigrated to Germany after the end of World War I;[12] and laws that expelled Jews from government employment[13] and permitted the "Aryanization" of the Jewish assets.[14] From such "hard-as-granite" laws Hitler and the German leadership fashioned their Final Solution.

Yet despite the enormity of these and later Nazi atrocities, the legal dilemma at Nuremberg was very real. In establishing a case against the defendants, especially on charges of "crimes against humanity," the prosecution had no pre-existing statutes sufficient to the task. However, the Allies did not hesitate to invoke the law—natural law—that supersedes all statutes and that abrogates the kind of perverted legal order such as existed in Nazi Germany. Peter Kreeft has written that the Nuremberg trial "assumed that such universal moral law really existed," whereas relativists such as Jean-Paul Sartre argued that it did not. Kreeft goes on to say that while Sartre's view may have been consistent with his existentialist philosophy it "was also moral bankruptcy—intolerable, unendurable, unlivable. Thank God for consistent moral relativists like Sartre: they show us what relativism really comes to."[15]

It is interesting that the standard used by the world to condemn the Nazis at Nuremberg was the same one Jefferson asked the "candid world" to use in vindicating America's Revolution. Had either of these judgments been based solely upon legal relativism, such vicious men as Hitler, Eichmann, and Himmler might have been considered law-abiding, and such virtuous men as Washington, Adams, and Jefferson might have been labeled outlaws. [16]

And the Nuremberg defendants and their attorneys argued strenuously that these were, after all, *honorable* men for whom duty was practically a sacred calling. As Herman Göering, founder of the Gestapo, put it: "I am here only to emphasize that I re-mained faithful to [Hitler], for I believe in keeping's one's oath not in good times only, but also in bad times when it is much more difficult." [17] Most of the other defendants made their cases in similar language. One exception was Hitler's chief architect and economic advisor, Albert Speer. Speer's attorney, Hans Flachsner, explained his client's conflict with Göering's assertion that they could not properly be tried for obeying their own laws:

> Speer felt, and argued very strongly from the moment . . . [the defendants] were allowed to associate, that a dictatorship readily makes and unmakes its laws, which therefore are no longer moral but political instruments. Because of this, he said, universal law, representing civilized thought, superseded national law. Under the universal law of civilization—which he said richly included pre-Hitler German civilization—they had to consider themselves responsible, and they should stand together in this court as honorable men and say so, loud and clear. [18]

But in this, and in Admiral Erich Raeder's assertion that theirs "was not guilt before a human criminal court but a crime before God," [19] it was assumed that although they might be *morally* guilty, they remained *legally* innocent; the defeated German state bore the guilt, they argued, not its people or its leaders.

To this Robert H. Jackson, an associate justice of the Supreme Court who served as America's chief counsel at Nuremberg, flatly stated: "We do not accept the paradox that legal responsibility should be least where the power is the greatest." Justice Jackson liked to recall English jurist Edward Coke's rebuke of James I's assertion of royal authority: "A King," Coke reminded the monarch, "is still under God and law." Jackson's natural-law reasoning was summarized by Ann and John Tusa, authors of *The Nuremberg Trial*:

> The very idea that states commit crimes, he said, "is a fiction. Crimes are always committed only by persons." Men who exercise great power cannot be allowed to shift their responsibility on to the fictional being, the State, "which cannot be produced for trial, cannot testify, and cannot be sentenced." He berated those who had sworn an oath of inviolable fidelity and absolute obedience to Hitler; he called it "an abdication of personal intelligence and moral responsibility." . . . Then came the cry from Jackson's heart, as characteristically honest as it was passionate. "I cannot subscribe to the perverted reasoning that society may advance and strengthen the rule of law by the expenditure of morally innocent lives but that progress in law may never be made at the price of morally guilty lives." [20]

Without the natural law, what defense do we have against the enfranchisement of evil? Can positive laws, which are those laws enacted by legislatures and interpreted by courts, ever absolve us of moral responsibility to the laws of nature? What relevance do the conflicting claims at Nuremberg have to the current American debate about "life" issues? To answer such questions we begin with a short sketch of the development of the natural-law tradition and its influence on American thought.

# 3

# Discovering Natural Law: An Historical Overview

*The idea of natural law is a heritage of Christian and classical thought. It does not go back to the philosophy of the eighteenth century, which more or less deformed it, but rather to Grotius, and before him to Suarez and Francisco de Vitoria; and further back to St. Thomas Aquinas; and still further back to St. Augustine and the Church fathers and St. Paul; and even further back to Cicero, to the Stoics, to the great moralists of antiquity and its great poets, particularly Sophocles. Antigone is the eternal heroine of natural law, which the Ancients called the unwritten law, and this is the name most befitting it.*

Jacques Maritain

11

NATURAL LAW is the *moral underpinning* of all man-made law. It is an unalterable, objective, universally binding, and eternally valid set of rules that can never be abrogated. It establishes the norms of morality without which we would be unable to distinguish right from wrong. All philosophical interpretations of natural law share a belief in the existence of certain fundamental legal principles and institutions which are in turn grounded in the plan of life—God's plan—that is inherent in all ordered social existence. Notre Dame law professor Charles Rice has called natural law "a set of manufacturer's directions written into our nature so that we can discover through reason how we ought to act."[1]

Reason is the key. So long as a person possesses this God-given ability, he or she can know the difference between good and evil and can act accordingly. This is why we speak of natural-law principles as "self-evident."

But just because they *may* be known does not mean that we know them. This paradox has something to do with lapses in contemporary education, something more to do with failures of modern morality, and a lot to do with human nature. And so it is important to articulate some examples of those laws that are "written in our hearts."

The standard formulation of natural law is this: do good and avoid evil. From culture to culture and from person to person variations may occur in what is meant by "good," but there will be utter consistency in the imperative to seek the good. According to Professor Rice, there are five basic, natural inclinations that we may know by the use of reason: to seek the good; to preserve oneself in existence; to preserve the species; to live in community; and to use intellect and will.[2] From these basic inclinations, man applies natural law by means of deduction and prudence.

And in no area of life is that deductive and prudential reasoning more critical than in the development of civil law. Throughout the ages, most legal scholars have accepted that conformity with natural law is the essential test of the ethical validity of any enactment or decision of a legislature or court. A man-made law that violates its natural-law antecedent is unjustifiable, and the willingness to ignore or even to sever this connection is what separates mere expediency from true justice. But, again, this does not mean that all civil or positive laws pertaining to a given legal issue ought to resemble one another in every particular. Indeed, civil law can be defined as the application of natural law principles to *changing* and *unique* conditions in society. As St. Thomas Aquinas (1225-1274) put it, the "general principles of the natural law cannot be applied to all men in the same way because of the great variety of human affairs."[3]

This "great variety of human affairs" is what makes the history of legal philosophy so fascinating, and the development of thinking about the natural law so intriguing. It is no exaggeration to say that political liberty and economic prosperity have flourished in direct proportion to the prominence of belief in natural law. "As a reminder to popular majorities and ruling politicians that their power is limited," Professor Rice writes, "natural law theory is always controversial and often unwelcome."[4] As the Nuremberg Trials demonstrated, the natural law is also indispensable. It is not a moldy relic of antiquity; it is not witchcraft; and it is not an exclusively Roman Catholic concern. In fact, the natural law was a focus of Greek philosophy more than a thousand years before Aquinas.

It was the view of Heraclitus (536-470 BC) that "all things flow, nothing abides," and this led him to conclude that there are eternal norms, universal laws that establish order for man. "Wisdom" he wrote, "is the foremost virtue, and wisdom consists in speaking the truth and in lending an ear to nature and acting according to him. . . . Wisdom is common to all. . . [A]ll human laws are fed by one divine law."[5]

In *Antigone*, one of the earliest tragedies of Sophocles (496-406 BC), the heroine defends her decision to violate the edict of King Creon prohibiting her from burying her dead brother:

> CREON (*To Antigone*): Now tell me, briefly and concisely: were you aware of the proclamation prohibiting those acts?
>
> ANTIGONE: I was. I couldn't avoid it when it was made public.
>
> CREON: You still dared break this law?
>
> ANTIGONE: Yes, because I did not believe that Zeus was the one who had proclaimed it, neither did Justice, or the gods of the dead whom Justice lives among. The laws they have made for men are well marked out. I didn't suppose your decree had strength enough, or you, who are human, to violate the lawful traditions the gods have not written merely, but made infallible. These laws are not for now or for yesterday, they are alive forever; and no one knows when they were shown to us first.[6]

Plato (427-348 BC) believed that there exists in nature an objective order, and that man can achieve virtue and happiness by following universal principles. In the *Laws*, he wrote "The unwritten laws of nature hold universally and underlie the civil law."

Aristotle (384-322 BC) recognized the need for civil law to be based on the natural law. In the *Nicomachean Ethics* he stated: "Of political justice part is natural, part legal—natural, that which everywhere has the same force and does not exist by people's thinking this or that; legal, that which is originally indifferent, but when it has been laid down is not indifferent."[7] In the *Rhe-*

*toric*, Aristotle expounded on this distinction: "[B]y general laws I mean those based upon nature. In fact there is a general idea of just and unjust in accordance with nature, as all men in a manner divine [i.e., know intuitively] even if there is neither communication nor agreement between them."[8]

What was true in Greece was true in Rome, only more so. Cicero (106-43 BC), the great philosopher-statesman, declared that the foundation of law is the *lex nata*, the divine and unchangeable law placed within us. In the *Commonwealth*, he wrote: "There is a true law, right reason [that is] in accord with nature; it is of universal application, unchanging and everlasting. . . . It is wrong to abrogate this law and it cannot be annulled. . . . There is one law, eternal and unchangeable, binding at all times upon all peoples; and there will be, as it were, one common master and ruler of men, God, who is the author of this law, its interpreter and its sponsor."[9]

For Cicero, justice could not be left to human law alone because it would lead to law that was based on the whims of those in control of government power:

> If the principles of Justice were founded on the decrees of peoples, the edicts of princes, or the decisions of judges, then Justice would sanction robbery and adultery and forgery of wills, in case these acts were approved by the votes or decrees of the populace. But if so great a power belongs to the decisions and decrees of fools that the laws of Nature can be changed by their votes, then why do they not ordain that what is bad and baneful shall be considered good and salutary? Or, if a law can make Justice out of Injustice, can it not also make good out of bad? But in fact we can perceive the difference between good laws and bad by referring them to no other standard than Nature: indeed, it is not merely Justice and Injustice which are distinguished by Nature, but also and without exception things

which are honorable and dishonorable. For since an intelligence common to us all makes things known to us and formulates them in our minds, honorable actions are ascribed by us to virtue, and dishonorable actions to vice; and only a mad-man would conclude that these judgments are matters of opinion, and not fixed by Nature.[10]

Seneca (3 BC-AD 65) and Epictetus (AD 50-138) both appealed to intrinsic human dignity and the natural-law basis of freedom and equality to condemn the practice of slavery.

At the beginning of the Christian era Saint Paul (died AD 67) taught that the natural law is in the hearts of all: Jew, Christian, and heathen. "Pagans who never heard the Law [of Moses]," he wrote in Romans 2:14-16, "but are led by reason to do what the Law commands, may not actually 'possess' the Law, but they can be said to 'be' the Law. They can point to the substance of the Law engraved on their hearts . . ."

The Fathers of the early Church also described the natural law as that which God gives to man and is known through the use of reason. Saint John Chrysostom (345-407) wrote:

We use not only Scripture but also reason in arguing against the pagans. What is their argument? They say they have no law of conscience, and that there is no law implanted by God in nature. My answer is to question them about their laws concerning marriage, homicide, wills, injuries to others, enacted by their legislators. Perhaps the living have learned from their fathers, and their fathers from their fathers and so on. But go back to the first legislator! From whom did he learn? Was it not by his own conscience and conviction? Nor can it be said that they heard Moses and the prophets, for Gentiles could not hear them. It is evident that they derived their laws from the law which God ingrafted in man from the beginning.[11]

Moses Maimonides (1135-1204) the renowned Jewish philosopher argues in *Guide for the Perplexed* that what is innate in man prescribes good actions and condemns bad ones; man is rewarded or punished accordingly, quite irrespective of the specific commands of a prophet.[12]

For Aquinas, man stands above all other creatures because of his power of reason, which permits man to know the natural law. According to Aquinas, "the natural law contains the precepts related to man's drive to preserve himself, which he shares with all things; those related to his animal drives such as sex and the rearing of children; and those which make him specifically human—his need for society, his desire for knowledge and for God."[13] Aquinas devoted a portion of the *Summa Theologica* (I-II, Questions 90-108) to the concept of the natural law and the objections of skeptics. "The first principle in the practical reason," Aquinas wrote, "is one founded on the nature of good, that good is that which all things seek after. . . . Hence this is the first principle of law, that good is to be done and pursued, and evil is to be avoided. All other precepts of the natural law are based upon this, so that whatever the practical reason naturally apprehends as good or evil belongs to the precepts of the natural law as something to be done or avoided."[14]

Concerning the natural law views of the great Protestant reformers such as Martin Luther (1483-1546), Philipp Melanchthon (1497-1560), Ulrich Zwingli (1484-1531) and John Calvin (1509-1564), Union Theological Seminary professor John T. McNeil has observed:

> There is no real discontinuity between the teaching of the Reformers and that of their predecessors with respect to natural law. Not one of the leaders of the Reformation assails the principle. Instead, with the possible exception of Zwingli, they all on occasion express a quite ungrudging

respect for the moral law naturally implanted in the human heart and seek to inculcate this attitude in their readers . . . [and] enters into the framework of their thought and is an assumption of their political and social teaching. . . . For the Reformers, as for the Fathers, canonists, and the Scholastics, natural law stood affirmed on the pages of Scripture.[15]

Writing in his *Commentaries on the Laws of England,* Sir William Blackstone (1723-1780) had this to say about the natural law:

The Creator . . . has so intimately connected, so inseparably interwoven the laws of eternal justice with the happiness of each individual that the latter cannot be attained but by observing the former. . . . In consequence of which mutual connection of justice and human felicity, he has not perplexed the law of nature with a multitude of abstracted rules and precepts . . . but has graciously reduced the rule of obedience to this one paternal precept, "that man should pursue his own true and substantial happiness." This is the foundation of what we call ethics or natural law. . . . This law of nature, being coeval with mankind and dictated by God himself, is superior in obligation to all others. . . . No human laws are of any validity if contrary to this.[16]

The founding fathers were influenced by Blackstone's *Commentaries* and by various British court opinions, including especially those of Edward Coke (1552-1634). According to philosopher Paul Sigmund:

Editions of Coke were very widely distributed in the colonies, and through them the colonists knew his decisions in *Calvin's Case* (1608) and in *Bonham's Case* (1609), which defended the superiority of the natural and common law to acts of Parliament and seemed to claim for the judiciary the right to enforce those limits on Parliamentary legislation. Coke's doctrines have been interpreted as stating a principle of judicial construction by a lower court of the law enunciated by the 'High Court of Parliament' rather than the assertion of a general right of judges to strike down parliamentary legislation. However, the plain words of the cases are "the common law will control the acts of Parliament and sometimes adjudge them to be utterly void" (*Bonham's Case*), and "the law of nature cannot be changed or taken away" and "should direct this case" (*Calvin's Case*).[17]

American statesman Alexander Hamilton (1757-1804) insisted that "no tribunal no codes, no systems can repeal or impair the laws of God, for by his eternal laws it is inherent in the nature of things,"[18] and in an 1850 speech, Senator Daniel Webster (1782-1852) cited the natural law in his denunciation of the expansion of slavery: "Now as to California and New Mexico, I hold slavery to be excluded from these territories by a law even superior to that which admits and sanctions it in Texas. I mean the law of nature. That law settles forever that slavery cannot exist in California or New Mexico."[19]

Following Webster's reasoning, Abraham Lincoln (1809-1865), in attempting to restore the Union and to obliterate slavery, consistently invoked the natural law in speeches and debates. In his famous oration at the Cooper Institute in 1860, he presented the natural-law argument with sublime subtlety:

> If slavery is right, all words, acts, laws, and constitutions against it are themselves wrong, and should be silenced, and swept away. If it is right, we cannot justly object to its nationality—its universality; if it is wrong, they cannot justly insist upon its extension—its enlargement. All they ask, we could readily grant, if we thought slavery right; all we ask, they could readily grant, if they thought it wrong. Their thinking it right, and our thinking it wrong, is the precise fact upon which depends the whole controversy.[20]

In modern times, journalist Walter Lippman (1889-1974) viewed the natural law as "the only conception of politics which is consistent with a free and civilized life."[21] And in his critically acclaimed work *The Moral Sense*, James Q. Wilson (1931- ) writes: "By a moral sense, I mean an intuitive or directly felt belief about how one ought to act when one is free to act voluntarily . . . [and] by ought I mean an obligation binding on all people similarly situated."[22]

And so from the noblest Greeks and Romans to the most saintly Christians and Jews, from men dead two-thousand years to men born in our century, the natural law has been advanced, articulated, and defended with eloquence and conviction. It has served as the basis of constitutional law in nations throughout recorded history, for without it a constitution is impossible. And without a constitution, the natural rights of human beings are little more than "values" competing with "interests."

# 4

# THE POLITICS OF NATURAL LAW:
# THE PERSON AND THE COMMON GOOD

*A person . . . is more than an individual. As the concept of* individual *looks to what is material, so the concept of* person *looks to intellect and will: the capacities of insight and judgment, on the one hand, and of choice and decision, on the other. A person is an individual able to inquire and to choose, and, therefore, both free and responsible.*

Michael Novak

BECAUSE MAN is a person who possesses a mind, he is substantially different from every other creature. Only man possess reason, imagination, creativity, and the capacities for moral and aesthetic thought. As the old saying has it: a man can make a monkey of himself, but no monkey can make a man of himself. It is man's mind, not his body, that is made the image and likeness of God and gives him his true dignity. Materialists may (apparently without irony) deny the existence of mind as a reality essentially different from matter, even though it obviously takes a mind to deny the existence of mind, since only a mind can affirm or deny anything. I would not be writing these words and you would not be reading them if there were no such thing as mind.[1]

Philosopher Mortimer Adler has written that human nature differs from other animal natures because only man possesses "the related powers of propositional speech and conceptual thought," and because his behavior is not governed by instinct. Man is free to choose. "He has, in short, the power of self-determination, the power of creating or forming himself and his life according to his own decisions."[2]

Man is an animal to be sure (and often acts like one), but he is a very special animal, and it is this uniqueness that is the foundation of the "inalienable rights" referred by Thomas Jefferson. Inalienable rights in the political sphere are rights which cannot justly be taken from citizens by the state, because the state did not grant these rights to the people to begin with. The people's inalienable rights come from God, the Author of human nature, and no fact of birth, wealth, or social position merits or diminishes them.[3] The liberties of the people are *natural* rights precisely because, as Jefferson put it, they "are the gift of God."[4] Jefferson's frequent antagonist and sometimes ally Alexander Hamilton agreed, describing inalienable rights as "written, as with a sunbeam, in the whole volume of human nature, by the hand of the Divinity Itself, and can never be erased or obscured by mortal power."[5]

Political man is a citizen (literally a "man of the city"), because by his very nature he is endowed with an appetite and inclination for social life. Nearly all people dislike solitude and seek companionship. They enjoy the company of others with whom they want to share their joys and sorrows and naturally come together socially, educationally, politically, and economically in communities. Aristotle pointed out that "the first thing to arise is the family . . . the association established by nature for the supply of men's everyday wants. . . . But when several families are united and the association aims at something more than supply of daily needs, the first society to be formed is the village . . . when several villages are united in a single complete community, large enough to be needy or quite self-sufficing, the state comes into existence. . . . Hence it is evident that the state is a creation of nature, and that man is by nature a political animal." [6]

We depend on one another: first upon our parents and then upon friends, neighbors, teachers, employers, etc. Individuals and families naturally broaden their associations to meet their mutual needs in the process called *subsidiarity*. The principle of subsidiarity affirms that decisions are most appropriately made by the local agencies closest to the relevant daily realities, and by the next highest agencies only when decisions and actions are beyond the capacities of those at lower levels. [7] The national government is the proper agency to wage war; the family is the proper agency to raise children. Subsidiarity calls for the highest feasible degree of self-autonomy: to the limits of his powers a person is to be entrusted with self-determination and the responsibility for the fulfillment of his own ends. Subsidiarity's emphasis on man's most significant freedom, freedom of the will, rests upon the conviction that man's faculties cannot be developed without his own effort. The greater the range of choices we face, the more self-reliance and personal responsibility are required. The more self-reliant and responsible we are the more we become what we are meant to be—and vice versa. Subsidiarity is the ultimate do-it-yourself principle. [8]

In his encyclical *Quadragesimo Anno* (1931), Pope Pius XI defined subsidiarity as "the fundamental principle of social philosophy, fixed and unchangeable, that one should not withdraw from individuals and commit to the community what they can accomplish by their own enterprise and industry."[9] Pope John XXIII re-enforced his concept in *Mater et Magistra* (1961):

> The state should leave to these smaller groups the settlement of business of minor importance. It will carry out with greater freedom, power, and success the tasks belonging to it, because it alone can effectively accomplish these, directing, watching, stimulating and restraining, as circumstances suggest or necessity demands. Let those in power, therefore be convinced that the more faithfully this principle be followed, and a graded hierarchical order exists between the various subsidiary organizations, the more excellent will be both the authority and the efficiency of the social organization as a whole and the happier and more prosperous the condition of the state.[10]

We unite with others in society in order to supply those things necessary for human development which neither the individual nor the family can provide. Our dependence on society becomes more acute as civilization becomes more complex. We need only experience a brief interruption in electric power to be reminded of our dependence upon society: no lamps to illuminate, televisions to entertain, or fans to cool. Such material needs are the bases of man's relation to and his dependence on others. Because we all need society, each of us becomes a part of the whole and is bound to society for the sake of the common good.[11]

The phrase, "the common good," is much maligned, in part because it is frequently used by people to rationalize and promote their special interests. Few would argue that the purpose of society and government is to achieve good, but the question remains, What kind of good? Individual, collective, or common?

The individual good benefits only one person. When I spend my money to purchase a meal in a restaurant, I am the only beneficiary. Starving people standing on the street outside do not benefit. If a government existed for the sole benefit of an autocrat, it could not possibly provide for the common good, although it might benefit the leader very nicely.

If one group benefits to the exclusion of other groups, it is a called a collective good. In Nazi Germany, for instance, the state benefited so-called Aryans and other members of the Nazi party while Jews, Gypsies, Catholics, and many others were excluded and even executed. Political systems based upon some version of collective good are little better for a whole society than those based upon individual good, and in some cases—Nazi Germany again—they can be worse.

Under a government dedicated to the common good, however, all members benefit. Possession or privilege by one person or group does not diminish or exclude possession or privilege by others. This inclusive character, called *distributive* good, is explicitly expressed in the words of the Pledge of Allegiance: "with liberty and justice for all." [12]

When we say the state exists for the "common good," two conclusions follow: a) the state does not exist for itself, which is to say for its ruling faction, but for the welfare of the people—no matter who they are; b) the common good is not to be considered in such a general way as to ignore the welfare of individuals—no matter who they are. [13]

But the person in not an atomistic individual, and the common good is superior to any private good. Thus in case of need the community may set aside the interests of individual citizens. The common good is therefore the good of the whole and of its parts, a good which subordinates man to society inasmuch as he is *social*, but also a good which respects man as a person ordered directly to God and his eternal end. Man is independent as regards his *value*—immortal, spiritual, and God-destined; he is dependent as regards his *function*—mortal, social, and bound to the common good. The individual functions somewhat like an organ in a body. Therefore man has both rights and duties. Society in like manner has rights and duties: rights as regards its end, which is the fostering of the common good; duties as regards its respect for inalienable rights which the state did not give and therefore cannot take away.[14]

In his classic work *The Person and the Common Good*, Jacques Maritain sums up the elements of the common good in society:

> [T]hat which constitutes the common good of political society is not only: the collection of public commodities and services—the roads, ports, schools, etc., which the organization of common life presupposes; a sound fiscal condition of the state and its military power; the body of just laws, good customs and wise institutions, which provide the nation with its structure; the heritage of its great historical remembrances, its symbols and its glories, its living traditions and cultural treasures. The common good includes all of these and something much more besides—something more profound, more concrete and more human. For it includes also, and above all, the whole sum itself of these; a sum which is quite different from a simple collection of juxtaposed units. (Even in the mathematical order, as Aristotle points out, 6 is not the same as 3 + 3.) It includes the sum or sociological integration of all the civic conscience, political virtues and sense of right and liberty, of all the activity, material prosperity and spiritual riches, of unconsciously operative hereditary wisdom, of moral rectitude, justice, friendship, happiness, virtue and heroism in the individual lives of its members. For these things all are, in a certain measure, *communicable* and so revert to each mem-

ber, helping him to perfect his life and liberty of person. They all constitute the good human life of the multitude.[15]

This discussion of the common good has described how man is naturally social and forms society by the demands and impulses of his rational nature working through his free will. Society is the enduring union of people morally bound under authority to cooperate for a common good. The end of society is the help which the individuals obtain from social cooperation as members of society for the fulfillment of their existential ends. Individuals fulfill their ends by their own activity. They are capable of doing so, however, only because their powers are complemented by social co-operation. Since this mutual aid is made available to all by the cooperation of all, it is called the common good.[16]

A natural society has a scope of authority, an extent of direction and compulsion, which the members as individuals never had or can have. Hence authority in a natural society cannot come from the individuals composing the society but must come from the author of the natural law from which natural societies derive their existence—God.

# 5

## NATURAL LAW UNDER FIRE: IDEOLOGICAL OPPONENTS OLD AND NEW

*Naturam expellas furca, tamen usque recurret.*

*[You may drive out nature with a pitchfork, yet it will always return.]*

Horace

FOR CENTURIES scores of ideologues have tried to refute or deny that moral absolutes are the basis of law. This section will describe the European strand of thought that systematically developed ways of denying, subverting, or avoiding natural-law considerations. Later sections examine another strand that has come to fruition—the American positivist-utilitarian notion of rights.

It was René Descartes (1596-1650), the father of modern philosophy, whose concept of man influenced European opposition to natural law. According to philosopher Larry Azar:

One important problem which Descartes bequeathed to modern thought stems from his dichotomy of the human person. Having firmly insisted that his essence was a mind (spiritual substance), he was logically forced to conclude that his body (matter) was *outside* his essence. Although he abhorred such a conclusion, it was inevitable from his having split man into two independent substances: If I am a mind (or soul), then my body is not part of my ego. This view contrasts with that of Aristotle, according to whom man is a psychosomatic composite: Man is body and soul. Here, the unity is in the man. Just as one molecule of a compound (e.g., table salt) is composed of different elements (e.g., sodium and chlorine), so too, a human being is, for Aristotle, composed of two parts: body and soul. For Descartes . . . the obvious unity of a person is lost.[1]

Azar concludes that Descartes' position "led to a materialism in which humans were to be reduced, literally to vegetable."[2] For Descartes' heirs, man is not a person, he is not substantially different from a beast, he is a mere individual differing only quantitatively from a rodent, an ape, or a plant.

Thomas Hobbes (1588-1679), the first modern advocate of the totalitarian state, built on this materialist reasoning and argued, first, that man does not possess the power of reason to discover the natural law, and, second, that man is not a social being and does not naturally form society. Hobbes contended that before man created, through a compact (or contract), the artificial device known as the state, he was antisocial and amoral. For Hobbes, "man was a wolf to man," and there was a "war of all against all." Man is ruled by his passions, and his life in the state of nature was "solitary, poor, nasty, brutish and short."[3] Man's natural rights did not stem from the natural law, because, in Hobbes' view, no such law exists. Man has liberty (license) to do anything he pleases in pursuit of self-preservation:

During the time men live without a common power to keep them all in awe, they are in the condition which is called war; and such a war as is of every man against every man...

To this war of every man against every man, this also is consequent: *that nothing can be unjust.* The notions of right and wrong, justice and injustice, have there no place. Where there is no common power, there is no law; where no law, no injustice. Force and fraud are in war the two cardinal virtues. Justice and injustice are none of the faculties neither of the body nor mind. If they were, they might be in a man that were alone in the world, as well as his senses and passions. They are qualities that relate to men in society, not in solitude. It is consequent also to the same condition, that there be no propriety, no dominion, no *mine* and *thine* distinct; but only that to be every man's that he can get; and for so long as he can keep it . . .

Where no covenant hath preceded, there hath no right been transferred, and every man has a right to everything; and consequently, no action can be unjust. . . . Before the names of just and unjust can have place, there must be some coercive power, to compel men equally to the performance of their covenants, by the terror of some punishment greater than the benefit they expect by the breach of their covenant . . . and such power there is none before the erection of a commonwealth.[4]

Man's impulse for self-preservation caused him to join with other men to end the constant warfare by implementing the social contract. Man gives absolute power to the sovereign, and the citizen's only rights are those granted by the head of state.

The state, according to Hobbes, was the "mortal god." In other words, there are no objective moral standards by which we may judge the laws of the state. Morality is defined by the state and is based on the whims and desires of the all-powerful sovereign. Accordingly, all law is public authority, the positive law of the state. Might makes right; law is will. Nothing can be called law which is not issued from a superior *visible* authority capable of enforcing its commands. Hence the natural law, which is not enforced as such by visible authority, should not be called law but moral principles—if reference is made to it at all.[5]

Jean Jacques Rousseau (1712-1778) was also an advocate of the social contract. For Rousseau, however, man in a state of nature lived in carefree innocence. The nation-state became essential because of the development of cultural mores that encouraged man to compete and acquire property. Thus did Rousseau seek to alter man's cultural environment through social engineering. Man is malleable, he believed, but to mold him requires a great and singular power to which Rousseau (who fancied himself capable of wielding the power to shape a new culture) gave the name "General Will." With power in hand, the benevolent state would proceed to define the structure of a new culture that would produce harmony among men. Government would define every aspect of man's activity and demand total submission. The dissenter would be eliminated. Historian Alexander Rustow viewed Rousseau's doctrine of the social contract as "nothing more than Hobbes' absolutist political theory, except that in place of the monarch he sets the equally absolute people."[6] As Rousseau himself put it: "Man must be forced to be free."

And in Rousseau's judgment, a natural law that defines absolute standards of morality does not exist. The social contract is the origin of morality and is entirely based upon convention:

> Man is born free, and everywhere he is in chains. Many a one believes himself the master of others, and yet he is a greater slave than they. How has this change come about? I do not know. What can render it legitimate? I believe that I can settle this question. . . . The social order is a

sacred right which serves as a foundation for all others. This right, however, does not come from nature. It is therefore based on conventions. The question is to know what these conventions are . . .

Since no man has any natural authority over his fellow-man, and since force is not the source of right, conventions remain as the basis of all lawful authority among men . . .

The passage from the state of nature to the civil state produces in man a very remarkable change, by substituting in his conduct justice for instinct, and by giving his actions the moral quality they previously lacked.[7]

The doctrines of Hobbes and Rousseau displayed little understanding of the graduated order of the forms of social life that resides in the nature of man as a social animal. They showed no appreciation for the family as a social institution with an essential end of its own. (They dealt only cursorily with marriage and parenthood.) They showed no concern for the occupational group or cooperative structure, and therefore gave no place in their scheme for the multifarious social forms that exist in all domains of life between the state and the individual. They certainly showed no regard for the principle of subsidiarity. In effect, Hobbes and Rousseau knew only the harsh antagonism of individual against state, a conflict which they believed the state must win.[8]

Philosopher David Hume (1711-1776) continued the assault on natural law. Man's nature is his passions. "Reason," wrote Hume, "is, and ought only to be the slave of the passions, and can never pretend to any other office than to serve and obey them."[9] To Hume the idea of objective moral principles is absurd. Such "principles" are only names and symbols for emotions; for feelings of pleasure and pain. What we call morals are nothing more than sentiments or habits. For Hume virtue is "to be whatever mental action or quality gives to a spectator the pleasure sentiment of approbation; and vice the contrary."[10]

In *A History of Political Theory*, author George Sabine impressively describes the implications of Hume's ideological approach: "If the premises of Hume's argument be granted, it can hardly be denied that he made a clean sweep of the whole rationalist philosophy of natural right, of self-evident truths, and of the laws of eternal and immutable morality which were supposed to guarantee the harmony of nature and the order of human society. . . . All the attempts to find in them an eternal fitness or rightness are merely confused ways of saying that they are useful; granted the principle of utility the whole system of natural right can be dispensed with."[11]

What was needed was a new set of morals devoid of law; a legal system devoid of morals. Although there is much of value in Hume's work (he was largely opposed to the rationalism of the age, and greatly influenced the work of Adam Smith) his utter skepticism helped give rise to utilitarianism and positivism.

Enter Jeremy Bentham (1748-1832). Bentham developed the philosophy known as *utilitarianism,* a moral system based on the so-called "pleasure principle."

Central to utilitarianism is the belief that the purpose of life, the highest good, is pleasure. Thanks to Bentham, hedonism has become respectable. He recycled the work of Epicurus (340-270 BC), who had described "good" as that which increases pleasure and bad as that which decreases it:

> We call pleasure the beginning and end of the blessed life. For we recognize pleasure as the first good innate in us, and from pleasure we begin every act of choice and avoidance, and to pleasure we return again, using the feeling as the standard by which we judge every good. And since pleasure is the first good and natural to us, for this very rea-

son we do not choose every pleasure, but sometimes we pass over many pleasures, when greater discomfort accrues to us as the result of them. . . . Every pleasure then because of its natural kinship to us is good, yet not every pleasure is to be chosen: even as every pain also is an evil, yet not all are always of a nature to be avoided. Yet by a scale of comparison and by the consideration of advantages and disadvantages we must form our judgment on all these matters . . .[12]

When Bentham took this concept and applied it to society, he came up with what he called the "Greatest Happiness Principle," which states that the good society is the one providing the greatest happiness to the greatest number. He wrote that "pleasure is in itself good, nay, even setting aside immunity from pain, the only good: pain is in itself evil; and indeed, without exception the only evil; or else [the] words good and evil have no meaning."[13]

In Bentham's view man consists of nothing more than matter; he has no "mind" and cannot know universal moral laws. He famously referred to the concept of natural law as "nonsense on stilts." Pleasure and pain are the only motives governing mankind.[14]

John Stuart Mill (1806-1873) a student of Bentham, agreed that "pleasure and freedom from pain are the only things desirable as ends," but Mill improved the theory by arguing that the greatest-happiness-for-the-greatest-number principle must be understood not merely by the quantity of pleasure but by its quality as well.[15] Mill introduced a governing principle that was to be the *sine qua non* of modern liberalism: "[The] only purpose for which power can be rightfully exercised over any member of a civilised community, against his will, is to prevent harm to others. His own good, either physical or moral, is not a significant warrant. He cannot rightfully be compelled to do or forbear because it will be better for him to do so, because it will make him happier, because, in the opinion of others, to do so would be wise or even right." [16] The common good is rejected in favor of the individual good.

When the moral basis of law is discarded, all that remains is the positive law. The good and the just are what is here and now deemed useful to the self-interest of individuals and to their life in common. Accordingly, law is the will of the state that is expressly declared to be such, is enacted in conformity with constitutional provisions, and is then duly promulgated. [17] Any further criterion, as, for example, the inherent justice or the moral lawfulness of the action commanded by the positive law, is rejected as irrelevant for the sphere of law. The legal sphere is solely identified with the state, its enforcement by the administration and the citizens and its application by judges. Law is consequently not normative; it barely pertains to reason. [18]

This approach was to have a tremendous impact on American constitutional law in the twentieth century.

6

# NATURAL LAW AND THE
# UNITED STATES OF AMERICA

*In drawing up the Constitution our forefathers did
not presume to establish a government which would
give individuals those rights which they had recognized and proclaimed as God-given. They merely
delineated and prescribed protections for those God-
given rights. They were quite familiar with theories
of law and government which made the state a God
and the subjects the mere recipients of privileges. They
would have none of it. To their way of thinking
deification of the state was absolutely antagonistic
to democracy. To their way of thinking God-given
rights were essential to democracy. To their way of
thinking there could be no democracy except on the
basis that authority to determine rights and duties
of individuals comes from God to the people and
through the people to those whom the people choose
to represent them. . . . For them there could be no
divorcing of law and morals. Morals was a broader*

36

*field, but the vital force of one was the vivifying,*
*preserving force of the other.*

Francis E. Lucey

T HROUGHOUT HIS LIFE, Thomas Jefferson espoused the necessity of recognizing man's social nature. Jefferson became increasingly critical of the modern liberal psychology which conceived of man as naturally individualistic, and of its corollary view of the state as merely a conventional artifice or "social contract" limited to preserving private rights of self-interest and self-preservation. Jefferson insisted that liberal natural-law philosophers, with ethical systems premised on man's selfish desires, should subordinate their theories to the realities of man's social nature which derived from the God-given moral sense.[1]

Jefferson was certainly influenced by John Locke (1632-1704) who subscribed to the social contract theory. But unlike Hobbes, Locke believed that man is naturally social and that a natural law exists that confers rights. Locke wrote: "God, having made man such a creature that in his own judgment it was not good for him to be alone, put him under strong obligation of necessity, convenience, and inclination to drive him into society, as well as fitted him with understanding and language to continue and enjoy it. The first society was between man and wife, which gave beginning to that between parents and children . . ."[2]

It is interesting to note that in rejecting Hobbes' antisocial view of man, Locke was indebted to philosopher Richard Hooker (1554-1600). In *Twentieth Century In Crisis: Foundations of Totalitarianism*, Larry Azar describes it this way: "Like Aquinas, Hooker discussed the various kinds of law: the eternal law, the natural law and the law of reason. . . . He reasoned that the political and social orders required natural law, and man's ultimate happiness necessitated divine law. For Hooker, then, men form society because they are *social* by *nature*. And it was from Hooker that John Locke borrowed his own doctrine that man is social by nature; and it was Locke who, through Montesquieu, indirectly whispered to Thomas Jefferson."[3]

Jefferson was also familiar with the writings of other proponents of the natural law. He read, among others, Robert Filmer, Algernon Sidney, Francisco Suarez, William Blackstone, and Edward Coke.

Writing to Francis Gilmer in 1816, Jefferson criticized Hobbes view of man as a "humiliation to human nature."[4] He described man's innate moral quality as "the brightest gem with which the human character is studded, and the want of it more degrading than the most hideous of bodily deformities."[5] Jefferson viewed "ethics as well as religion as supplements to law in the government of men" and that "the state's moral rule of [its] . . . citizens" will be enhanced by its "enforcing moral duties and restraining vice."[6]

In that same year Jefferson wrote to John Adams that man is "an animal destined to live in society" because "the Creator . . . intended man for a social animal."[7] On another occasion he wrote: "The Creator would have been a bungling artist, had he intended man for a social animal, without planting in him social dispositions."[8]

The author of the Declaration of Independence was not the only founding Father to view the natural law as the fulcrum of society. Alexander Hamilton wrote this defense of the legality of actions by the Continental Congress:

> There are some events in society to which human laws cannot extend, but when applied to them lose their force and efficacy. In short when human laws contradict or discountenance the means which are necessary to preserve the es-

sential rights of any society, they defeat the proper end of
all laws and so become null and void. . . . The sacred rights
of mankind are not to be rummaged for among old parch-
ments or musty records. *They are written as with a sunbeam
in the whole volume of human nature, by the hand of Divinity
itself and can never be erased or obscured by mortal power.*[9][Italics
added.]

For Hamilton "no tribunal, no codes, no system can repeal or
impair the law of God, for by his eternal laws it is inherent in
the nature of things."[10]

George Mason's *Declaration of Rights* (1787), which was
adopted as the preamble of the Virginia Constitution, refers to
the natural law and to the common good:

1. That all men are by nature equally free and independent,
   and have certain *inherent right*, of which, when they enter into a state of society, they cannot, by any compact,
   deprive or divest their posterity; namely, the enjoyment
   of life and liberty, with the means of acquiring and possessing property, and pursuing and obtaining happiness
   and safety.

2. That all power is vested in, and consequently derived from,
   the people; that magistrates are their trustees and ser-
   vants, and at all times amenable to them.

3. That government is, or ought to be, instituted for the
   *common benefit*, protection, and security of the people, nation, or community. . . .[11] [Italics added.]

At his first inauguration (1789), George Washington declared
"it would be peculiarly improper to omit in this first official Act
my fervent supplications to the Almighty Being who rules over
the Universe, who presides in the Councils of Nations, and whose
providential aids can supply every human defect . . ."[12] In his
farewell address, Washington reminded the nation that "of all

the dispositions and habits which lead to political prosperity, religion and morality are indispensable supports. . . . Reason and experience both forbid us to expect that national morality can prevail in exclusion of religious principle."[13] John Adams agreed. "Our Constitution," he wrote, "was made only for a religious and moral people. It is wholly inadequate for the government of any other."[14]

Natural-law doctrines of limited government and inalienable rights prepared the way for the institution of judicial review, and it was by an appeal to the nature of a constitution as fundamental law that John Marshall, in *Marbury v. Madison* (1803), claimed for the Supreme Court the right to decide on the constitutionality of legislative acts. However, natural-law thinking was evident in Supreme Court decisions well before the *Marbury* case. In *Chisholm v. Georgia* (1793), Chief Justice John Jay appealed to "reason and the nature of things" to justify suits by a private citizen against a state. In *Calder v. Bull* (1798), Justice Samuel Chase spoke of certain "vital principles of our free government," including "the great first principles of the social compact" (among them private property), and "the general principles of law and reason," which would forbid the enactment of *ex post facto* legislation even if there were no constitutional prohibition.[15]

The opponents of slavery invoked the natural law. As early as 1772, George Mason made his position clear when, before the General Court of Virginia, he argued against a slavery statue:

> All acts of legislature apparently contrary to natural right and justice are, in our laws, and must be in the nature of things, considered as void. *The laws of nature are the laws of God; Whose authority can be superseded by no power on earth.* A legislature must not obstruct our obedience to him from whose punishments they cannot protect us. All human

constitutions which contradict his laws, we are in conscience bound to disobey. Such have been the adjudications of our courts of Justice.[16] [Italics added.]

And as mentioned above, both Daniel Webster and Abraham Lincoln cited natural law in their attacks upon slavery.[17]

It should be obvious that for the Founding Fathers and their nineteenth-century heirs, man, the person, had intrinsic value. The cornerstone of American democracy is the concept of the person: of his dignity, and of his inalienable rights, duties, and freedoms.[18] In the twentieth century, however, the tendency in American law has been to dismiss this concept of man. America has witnessed the rise of philosophical pragmatists and legal realists for whom natural law and its relation to the person is anathema.

# 7

# AMERICAN IDEOLOGICAL OPPOSITION TO NATURAL LAW

*Man has no dignity if he is not a rational animal,
essentially distinct from the brutes by reason of the
spiritual dimension of man.*

Moritmer Adler

WRITING TO HIS BROTHER Henry, in May 1907, William James (1842-1910) boasted about the impact he expected his 1906 lectures, published in *Pragmatism*, to have on American thought: "I shouldn't be surprised if ten years hence it should be rated as 'epoch making', for of the definitive triumph of that general way of thinking I can entertain no doubt whatever—I believe it to be something quite like the Protestant reformation."[1]

The elder James' prediction has in many ways come to pass. It was his philosophy, known as pragmatism, that was to begin the erosion of natural-law influence in American jurisprudence. In fact it led to contempt for the natural law.

In 1906 James proclaimed that a "pragmatist turns his back resolutely and once for all upon a lot of inveterate habits dear to professional philosophers. He turns away from abstraction and insufficiency, from verbal solutions, from bad *a priori reasons, from fixed principles, closed systems, and pretended absolutes and origins.* He turns toward concreteness and adequacy, towards facts, towards action and towards power."[2]

Theories for James became nothing more than instruments: "[Pragmatism] agrees with nominalism . . . in always appealing to particulars; with utilitarianism in emphasizing practical aspects; with positivism in its disdain for verbal solutions, useless questions and metaphysical abstractions."[3]

Ideas and opinions were considered true so long as they had a utilitarian value; so long as they are useful:

> A new opinion counts as "true" just in proportion as it gratifies the individual's desire to assimilate the novel in his experience to his beliefs in stock. It must both lean on old truth and grasp new fact; and its success . . . in doing this, is a matter for the individual's appreciation. When old truth grows, then, by new truth's addition, it is for subjective reasons. . . .[4]
>
> That new idea is truest which performs most felicitously its function of satisfying our double urgency. It makes itself true, gets itself classed as true, by the way it works.... *Purely objective truth . . . is nowhere to be found.*[5] [Italics added.]

For James, "an idea is true so long as to believe it is profitable to our lives." The truth is only an expedient. It is relative:

> Expedient in almost any fashion; and expedient in the long run and on the whole, of course; for what meets expediently all the experience in sight won't necessarily meet all farther experiences equally satisfactorily. . . . [W]e have to

live today by what truth we can get today and be ready tomorrow to call it falsehood. Ptolemaic astronomy, Euclidian space, Aristotelian logic, scholastic metaphysics, were expedient for centuries, but human experience has boiled over those limits, and we now call these things *only relatively true*, or true within those borders of experience. "Absolutely" they are false.[6]

The key to understanding James is to grasp that there are no absolute truths; no unconditional principles or laws. All is relative. The basis of democracy is the ever changing push and pull of diverse opinions and tastes. Skepticism rules; man must act on his impulses. Indeed action is all that matters. The history of philosophy, James wrote, pillaging Shakespeare, is "full of sound and fury signifying nothing." Hence the former beliefs in transcendent order, metaphysics, common law, and other customs and prescriptions must be replaced with concepts that are workable, efficient, and materialistic. Laissez-faire individualism and the survival of the fittest replaced the common good.

James was not alone in his rejection of belief in the divine nature of man and society. In 1906 William Graham Sumner published *Folkways* in which he re-enforced the pragmatism of James:

> Men begin with acts, not with thoughts. . . . The ability to distinguish between pleasure and pain is the only psychical power which is to be assumed. Thus ways of doing things were selected, which were expedient.
>
> The folkways . . . are not creations of human purpose and wit . . . they are like the instinctive ways of animals, which are developed out of experience. . . .
>
> The notion of right is in the folkways. It is not outside of them. . . . In the folkways, whatever is, is right. . . "Rights" are the rules of mutual give and take in the competition of life which are imposed on comrades in the in-

group. . . . *Therefore rights can never be "natural" or God-given," or absolute in any sense.* . . . World philosophy, life policy, right, rights, and morality are all products of the folkways.

[T]he great mass of any society lives a purely instinctive life just like animals. . . . The masses are the real bearers of the mores of the society. . . . The folkways are their ways. . . . Institutions and laws are produced out of mores.

Nothing but might has ever made right. . . . If a thing has been done and is established by force . . . it is right in the only sense we know. . . .[M]ight has made all the right which ever has existed or exists now.[7] [Italics added.]

Influenced by Darwin's theory of evolution, the pragmatists maintained that man was not made in the image and likeness of God, and thus that there is no difference between man and beast. Since war is natural among beasts, only the fittest economically or politically will survive. As Sumner said: might makes right.

Another pragmatist, John Dewey, savoring the perceived victory of his ideological school of thought, declared that the "intellectual basis of the legal theory of natural law and natural rights had been undermined by historical and philosophical criticism."[8] If there is no natural law, no objective moral law, then there can only be positive law. The classic definition of positive law was articulated by John Austin, a student of utilitarian Jeremy Bentham: "Law consists of nothing more nor less than commands backed by threats of force and issued by superiors to inferiors who are in the habit of obeying."[9]

In the history of American law the foremost proponent of positive law and the philosophy of legal realism was Oliver Wendell Holmes (1841-1935), perhaps the most dominant force ever to sit on the United States Supreme Court. Felix Frankfurter proclaimed that Holmes "stood above all others, [and] has given the directions of contemporary jurisprudence." [10] For Franklin Roosevelt, Dean Acheson, Alger Hiss and countless others of their generation, Holmes was *sui generis* in thought and action.

The pragmatism of James and Dewey and the theories of Charles Darwin had a tremendous impact on Holmes' philosophy of law. Holmes was one of those who "never had an unpublished thought," and he left a voluminous collection of letters and essays that spell out his ideological perspectives. From the mass of material he left behind, not least from his Court opinions, we can study his views on God, human nature, society, moral law, and rights, and begin to sketch his concept of civil society.

Oliver Wendell Holmes was an atheist for whom the world is at best nothing more than a congeries of evolving energies. "I think the proper attitude," he wrote to Sir Frederick Pollock, "is that we know nothing of cosmic values and bow our heads— seeing reason enough for doing all we can and not demanding the plan of a campaign of the General—or even asking whether there is any general or any plan." [11]

Holmes was a complete skeptic. He believed that nothing can be known with certainty. He even questioned his own existence:

> I noticed once that you treated it as a joke when I asked how you knew that you weren't dreaming me. I am quite serious, and as I have put it in an article referred to above, we begin with an act of faith, with deciding that we are not God, for if we were dreaming the universe we should be God so far as we knew. You never can prove that you are awake. By an act of faith I assume that you exist in the same sense that I do and by the same act assume that I am in the universe and not it in me. I regard myself as a cosmic ganglion—a part of an unimaginable, and don't

venture to assume that my *can't helps,* which I call reason and truth, are cosmic *can't helps.*[12]

Although there are no certainties, Holmes took it on a kind of faith that objects did exist around him. He could not help feeling this way but asserted that the way he felt may not be real outside of his mind or imagination. His "can't helps" were those things in experience which he knew but could not confirm as real. They are valid "for my world," he wrote, "but which I can't assert concerning the world, if there is one."[13] The mind cannot know objective values, and all a man can do is *bet* on existence, on his surroundings, and try to predict the direction of forces around him: "Chauncey Wright, a nearly forgotten philosopher of real merit, taught me when young that I must not say *necessary* about the universe, that we don't know whether anything is necessary or not. So I describe myself as a *bet*-tabilitarian. I believe that we can *bet* on the behavior of the universe in its contact with us. We bet we can know what it will be. That leaves a loophole for free will—in the miraculous sense—the creation of a new atom of force although I don't in the least believe in it."[14]

Based on this epistemology, Holmes concluded that absolute truths and, consequently, all religions are illusions. If there is no God, then there is no design, and man cannot really possess objective values. There can be no judgments that are universal and necessary; truth is a chimera, merely one's perceptions based on one's abilities. In an article called "Ideals and Doubts" he defined truth "as the system of my limitations, and leave absolute truth for those who are better equipped. With absolute truth I leave absolute ideals of conduct equally on one side."[15] Holmes also stated that "truth was a majority vote of the nation that could lick all others;"[16] not absolute, but relative to time and place. All ideas, concepts, values are tentative.

There is no substantial difference in the world Holmes imagined between man and ape, who, as Descartes believed, differ only in degree. Man, therefore, has no inalienable rights. He is stripped of all dignity, acting purely by instinct or stimulus and response, and awaits the civilizing force of positive law. "I don't believe," Holmes wrote, "that it is an absolute principal or even a human ultimate that man always is an end in itself—that his dignity must be respected, etc. . . ."[17] In one letter he wrote: "I hardly think of man as so sacred as [Harold] Laski seems to think . . . I shall think socialism begins to be entitled to serious treatment when and not before it takes life in hand and prevents the continuance of the unfit."[18] There is finally only the struggle to survive:

> The struggle for life, undoubtedly, is constantly putting the interests of men at variance with those of the lower animals. And the struggle does not stop in the ascending scale with the monkeys, but is equally the law of human existence. Outside of legislation this is undeniable. It is mitigated by sympathy, prudence, and all the social and moral qualities. But in the last resort a man rightly prefers his own interest to that of his neighbors.[19]

> For my own part, I believe that the struggle for life is the order of the world, at which it is vain to repine. But I know of no true measure of men except the total of human energy which they embody. . . . The final test of this energy is battle in some form.[20]

In Holmes' Godless world there can be no natural law, no first principal: "The jurist who believes in natural law seems to me to be in a naive sense of mind."[21] There is only legal positivism (a.k.a. legal realism, analytical jurisprudence)—authority based on force. Thomas Holland, agreeing with Holmes, wrote in his work *Jurisprudence*: ". . . that which gives validity to a legal right is, in every case, the force which is lent to it by the state. Anything else may be the occasion, but is not the cause, of its obligatory character."[22]

With no appeal to absolute values there are no "oughts," and, as a consequence, the dominant group in power has no limitations. Those with power implement whatever has economic, social, or political utility. There are no transcendent standards of morality, no connection between moral absolutes and law, no bad conduct, no sin. Morals for Holmes are only emotional tasks:

> Our system of morality is a body of imperfect social generalizations expressed in terms of emotions.[23]

> I should be glad if we could get rid of the whole moral phraseology which I think has tended to distort the law. in fact even in the domain of morals I think that it would be a gain, at least for the educated, to get rid of the word and notion Sin.[24]

> It would be well if the intelligent classes could forget the word sin and think less of being good. We learn how to behave as lawyers, soldiers, merchants, or what not by being there. Life, not the parson, teaches conduct.[25]

One cannot even say that murder is inherently wrong; it is merely disliked by those who dominate society.

Authority is the applied physical force of a majority (in a democracy) or of whatever force dominates society. Public policy is whatever that dominant force wants, even if it is mob violence. Holmes said: "I am so skeptical as to our knowledge about the goodness or badness of laws that I have no practical criticism except what the crowd wants."[26] The basis for civil society is not men as social beings coming together naturally but force alone: "I believe that force, mitigated so far as may be by good manners, is the *ultima ratio*, and between two groups that want to make inconsistent kinds of world I see no remedy except force. I may add what I no doubt have said often enough, that it seems to me that every society rests on the death of men. . . ." [27]

Holmes explains that civil society is formed because men, like beasts, often have similar interests. It is a dictatorship of force among dominant elites who coerce the compliance of those others who would be members of society: "If I do live with others they tell me that I must do and abstain from doing various things or they will put the screws on to me. I believe that they will, and being of the same mind as to their conduct I not only accept the rules but come in time to accept them with sympathy and emotional affirmation and begin to talk about duties and rights." [28] Only superior force makes everything work; a majority's feelings determines truth, morals, and law. Public policy varies with the tastes of those in power:

> So when it comes to the development of a *corpus juris* the ultimate question is what do the dominant forces of the community want and do they want it hard enough to disregard whatever inhibitions may stand in the way. If a given community has a definite ideal, for instance, to regulate itself so as to produce a certain type of man, other communities would have a different one—and that community might change in a hundred years. But suppose the ideal accepted, there would be infinite differences of opinion as to the way in which it was to be achieved, and the law of a given moment would represent only the dominant will of the moment, subject to change on experiment or for deeper reasons. But I am beginning too far back. You assume a body of law is in force and start to formulate the principles of juristic development. I should think the only principles worth talking about were the existing notions of public policy. [29]

From the authority of those fit to survive (the authoritarian majority) legislative law is defined: The more powerful interest must be more or less reflected in legislation; which like every other device of man or beast, must tend in the long run to aid the survival of the fittest.[30]

Legislative Law is dominated by the supreme power. In this might-makes-right environment, the survival of minorities depends on the sympathies of those in power. "All that can be expected from modern improvements," Holmes wrote, "is that legislation should easily and quickly, yet not too quickly, modify itself in accordance with the will of the *de facto* supreme power in the community, and that the spread of an educated sympathy should reduce the sacrifice of minorities to a minimum."[31]

Rights and duties do evolve in Holmes' society, but how he defines them is revealing. In a letter to Pollack, Holmes asks, What is a right?

> . . . It starts from my definition of law (in the sense in which it is pursued by the modern lawyer), as a statement of the circumstances in which the public force will be brought to bear upon men through the courts: that is the prophecy in general terms. Of course the prophecy becomes more specific to define a right. So we prophesy that the earth and sun will act towards each other in a certain way. Then as we pretend to account for that mode of action by the hypothetical cause, the force of gravitation, which is merely the hypostasis of the prophesied fact and an empty phrase. So we get up the empty substratum, a *right*, to pretend to account for the fact that the courts will act in a certain way.[32]

Rights are illusions. What we call a "right" is only a person's rationale to obey force:

> I can imagine a book on the law, getting rid of all talk of duties and rights—beginning with the definition of law in the lawyer's sense as a statement of the circumstances in which the public force will be brought to bear upon a man

through the Courts, and expounding rights as the hypostasis of the prophecy.[33]

The law talks about rights, and duties, and malice, and intent, and negligence, and so forth, and nothing is easier, or, I may say, more common in legal reasoning, than to take these words in their moral sense, at some stage of the argument, and so to drop into fallacy.[34]

The Constitution, for Holmes, is only an experiment: "the claim of our especial code to respect [the Constitution] is simply that it exists, that it is the one to which we have become accustomed, and not that it represents an eternal principle."[35] Therefore, the most the Supreme Court can do is predict how tastes will change: "The object of our study, then, is prediction of the incidence of the public force through the instrumentality of the courts . . . a legal duty so called is nothing but a prediction that if a man does or omits certain things he will be made to suffer in this or that way by judgment of the court; and so of a legal right."[36]

Judicial decisions reflect nothing more than the workings of the moment. They have no ultimate or absolute value beyond "what is" at any moment in time. There are no ultimate ends because that would infer an objective norm. Law is only descriptive, it describes what has happened, not what should happen. Seeking precedents in case law is absurd because tastes change as the forces that dominate the social and political environment change. For Holmes and his disciples a judicial decision is "a social event. Like the enactment of a federal statue, or the equipping of a police car with radios, a judicial decision is an intersection of social forces."[37]

According to Holmes the "administration of justice" is also a subjective matter. The United States does not have laws governing men, because men govern men. As one legal realist put it: "Let us banish from our professional tenets the absurd dogma 'a government of laws not of men . . .' there is no place for it in legal science."[38] If men only avoid committing crimes, it is not because crime is morally wrong but out of fear that those who dominate society will punish them.

Needless to say, this approach to the law did not die with Holmes; if anything, it has become the dominant view, at least in the law schools of elite universities. One contemporary legal scholar, Ronald Dworkin, suggests in *Law's Empire* (1986) that the purpose of law "is to legitimate the community's exercise of force."[39] Another well-known scholar, John Rawls, has argued—in *A Theory of Justice* (1971)—that rational decision makers must be "situated behind a veil of ignorance."[40] "Rawls tries to convince us," writes philosopher J. Budziszewski, "that in order to figure out the principles of justice we must pretend to forget not only who we are but also everything we ever thought we knew about morality."[41] Harvard Law Professor Laurence Tribe, who argued in favor of the "right to die" before the U.S. Supreme Court in January, 1997, questioned the fitness of Clarence Thomas to sit on the Supreme Court because of his "adherence to 'natural law' as a judicial philosophy." Tribe concluded that Thomas' view "could take the Court in an even more troubling direction."[42]

These and others of the followers of Oliver Wendell Holmes are espousing a philosophy of expediency. They are driven to transform American culture according to their current view of its best interests, and the means for the changes they would effect is raw judicial power. They are crypto-totalitarians who deny the intrinsic value of man and seek total control over him. To them liberty means obedience to the uncertain will of the legal elite. To them the state, not God, is absolute.

To prevent a Holmesian dictatorship, America needs the natural law.

8

# The Need for Natural Law

*[W]e have arrived at the point historically where we can no longer proceed with any health or happiness on the blithe assumption that it doesn't matter what any of us believe—or whether there is really anything to believe. I submit to you today that we ought to believe what is true, and that the truth is that we live in a moral universe, that the laws of this country and of any country are invalid and will be in fact inoperative except as they conform to a moral order which is universal in time and space. Holmes held that what I have just said is untrue, irrelevant, and even dangerous.*

Henry R. Luce, 1951

JUSTICE HOLMES WROTE: "Theory is the most important part of the dogma of the law, as the architect is the most important man who takes part in the building of a house. . . . It is not to be feared as unpractical, for, to the competent, it simply means

54

going to the bottom of the subject."[1] Mr. Justice Holmes is correct in part; every body of law presupposes a philosophy.

If that philosophy subscribes to the belief that a Creator endowed men with inalienable rights, and that resting on a divine foundation the state is not autonomous but subject to a higher law, then judges and legislators will respect the inherent dignity of men by recognizing that government's powers are limited. God as the source of law becomes an animating principle of democracy, while the method of democracy becomes the process by which, for the common good, the executive, legislative, and judicial branches of government order the people's God-given rights.

If, however, the underlying philosophy assumes that morality is divorced from law and the force that controls state power grants men their rights (and may manipulate or withhold the rights of life, liberty, and property at any time), then men become mere instruments of those who wield power and are compelled to endure any indignities.

In the Post-World War II era, the underlying philosophy that has made disastrous inroads into our academic, governmental, and legal institutions is positivism, which has been defined as the "influential view (a version of rationalism) that only science provides real knowledge of the world, and that accordingly rejects all religious and metaphysical explanations of human experience."[2] Consequently, the idea of an objective moral order based on a synthesis of reason and Revelation is considered the outdated residue of medieval times: in the lexicon of the logical positivists it is a pseudo-problem. For the legal positivist, only human law is valid. Today, legal scholar Charles Rice tells us, "The legislator decides what law will be useful and in accordance with the basic norms as determined by himself. Once a law is enacted it is obligatory. There is no higher law of nature or of God, and the ultimate criteria is force."[3]

This "legal realism" has so infiltrated the country that Chief Justice Fred Vinson wrote in *Dennis v. United States* (1951): "Nothing is more certain in modern society than the principle that there are no absolutes . . . we must reply that all concepts are relative."[4]

Reacting to that declaration, Felix Morley wrote in the June 1951 issue of *Barron's* that "it becomes definitely dangerous to the spiritual welfare of the Republic when the chief law officer of its Government declares irrelevantly and even irreverently, that 'all concepts are relative.'. . . It stems back to Justice Oliver Wendell Holmes . . ."[5]

How, we may ask, does American legal positivism differ from the ideology that dominated the Nazi system? A German decree of 1936 stated that a decision of the Führer in the express form of a law or decree may not be scrutinized. The American judicial system, dominated by the epigones of Holmes, believes that nine "führers" determine the law of the land.

Hitler, like Holmes, was a Godless Darwinist. He believed only the fit had the right to survive. "The great masses," Hitler wrote "are only a part of nature. . . . What they want is the victory of the stronger and the annihilation or the unconditional surrender of the weaker."[6] The so-called rules of humanity do not apply to man; "[he] lives or is able to preserve himself above the animal world, but sadly by means of the most brutal struggle."[7]

Like Holmes, Hitler rejected God and the natural law; for him "the racial community is the basis and the source of the law."[8] All struggles occur within the race. Therefore, man has no dignity, no inalienable rights, he is a beast: "The main plank of the National Socialist program," Hitler declared, "is to abolish the liberalistic concept of the individual . . ."[9] Man was not exempt from the laws that govern the beast: "[Man] must understand the fundamental necessity of Nature's rule and realize how much his existence is subjected to these laws of eternal fight and upward struggle. . . . *There can be no special laws for man.*"[10] Based on this view of man's nature, Hitler's conclusion was similar to Holmes: "Only force rules. Force is the first law." Only "the stronger man is right."[11]

Hitler, like Holmes, argued that government policy should be based on the "heroic will." "The essential thing," according to Hitler, "is the formation of the political will of the nation."[12] That will, however, was to be in the hands of one man: the Führer. Göering pointed out that: "The law and the will of the Führer are one." Hans Frank stated: "[O]ur constitution is the will of the Führer." And Hitler himself declared to his nation: "My will decides;" "The fate of the Reich depends on me alone;" "There must be no majority decision, the decision will be made by one man."[13] Hitler instructed the German masses that the "will of the Nation [Hitler's will] is to be valued more highly than the individual's freedom of mind and will."[14]

Speaking before the March 1941 annual Washington meeting of the National Social Science Honor Society, Francis Lucey, Regent of Georgetown University's Law School and Professor of Philosophy and Jurisprudence, pointed out to his audience what was wrong with the Nazi philosophy: "It is the false philosophy of life, law and government; that philosophy of healthy, warm, well-sheltered and well-trained animals, mighty lions with the dignity and independence that a cage provides, a philosophy that divorces itself from the moral dignity of man. Why did the Nazis adopt this system? Because their philosophy is the philosophy of the deification of the state and social interests, claims and demand, *and the philosophy of what is useful and works; realism, utilitarianism and pragmatism.*[15]

The pragmatism and utilitarianism of Hitler is very similar to Holmes' philosophy. "The difference between Holmes and Hitler," Lucey pointed out, "was that Hitler decided that the elite was not always changing but instead was limited to a particular race. It is true that Holmes was a gentlemanly Darwinian. He too would like, as he often repeated, to get rid of the unfit but he would be more apt to do it by sterilization or gas chambers than with a butcher's cleaver. His was a kid-glove Darwinism." [16] There is also one other difference, Hitler had the opportunity to demonstrate the effects of the implementation of the pragmatist approach. Force, struggle, the will of the dominant power led to the annihilation of millions of innocent people.

In November 1945, Law Professor Ben W. Palmer published an essay in the *American Bar Association Journal*, that rocked the foundations of the American legal profession. Entitled *Hobbes, Hitler and Holmes,* it stated: "[Holmes'] basic principles lead straight to the abasement of man before the absolutist state and the enthronement of a legal autocrat—whether individual, minority or majority—a legal autocrat who may perhaps be genial as Holmes, benevolently paternalistic, perhaps grim and brutal as any Nazi or Japanese totalitarian, but nonetheless, an autocrat in lineal succession from Caesar Augustus on Nero through Hobbes and Austin and Mr. Justice Holmes." [17]

Based on his philosophy, Holmes could not object to the legality of the Nazi government's utilization of power—the most he could say is that his tastes were different. One intellectual heir of Holmes, Hans Kelsen, put it this way: "The legal order of totalitarian states authorizes their Governments to confine in concentration camps persons whose opinions, religion or race they do not like; to force them to perform any kind of labor, even to kill them. Such measures may be morally or violently condemned; but they cannot be considered as taking place outside the legal order of those states. . . . From the point of view of the science of law, the law under the Nazi government was law. We may regret it but we cannot deny that it was law." [18]

Holmes and his Darwinian followers might have explained the German incident pragmatically as follows: Hitler's tastes just differed from ours. While Hitler and his followers had the power, whatever they ordered was good and true. When we won the war, we had the power, and our tastes became the good and the true. Since we did not like their taste and had the power in our hands, we killed them, but it was just a matter of taste and power. The cards in the hands of the dominant players are always trump cards.[19] Or so say the Holmesians.

What are the ultimate dangers to our democracy if these legal realists triumph and the only standard of value is "might makes right?" Lucey, writing in 1942, explained it best:

> It would destroy Democracy, because the essential tenets of Realism contradict all the fundamental principles of Democracy. Democracy demands a union of men who are *de jure* independent by the nature God gave them. Realism denies that *de jure* independence, and reduces Democracy to a union of superior house-broken animals, or ganglia, who live by privilege, granted to them by the dominant physical forces in the union. Democracy demands government by all the people. Realism demands government by some, the dominant. Democracy rejects the tenet that might or physical force constitutes authority, the right and the duty. Democracy demands that even a minority have some inalienable rights. Realism has no protection for a minority. The minority is entirely and completely at the mercy of the dominant group. Democracy has a definite permanent end, the good of all the people. Realism has no end, unless we call the completely changeable pragmatic pattern of the dominant, an end. Democracy with its definite objective or end, with its established norms, respect for man's inalienable rights, respect for minority rights, supremacy of reason and objective values, promotes, despite divergent views and changing conditions, a sense of security, unity and happiness. Realism, by reason of its instrumental "is," its rejection of the role of reason, its relativism, and its skep-

ticism of values, could promote nothing but doubt, confusion, division, fear and despair, the playground in any
nation for demagogues, sophists and dictators. . . . Applied
Realism would have to destroy both the Constitution (which,
of course, can be done by interpreting it out of existence)
and the conviction of the people. Applied Realism in short
would destroy Democracy.[20]

Throughout this Century, American's law schools, legal jour-
nals, bar associations, legislative halls and court rooms have been
innundated with legal realism. And this philosophy of jurisprudence has resulted in startling constitutional interpretations.

In 1970, for instance, the New York State Court of Appeals,
in *Byrn v. New York City Health & Hospital Corporation*, ruled
that the state legislature, not God, has the power to determine
who is a person entitled to rights. As authority, the court quoted
the noted legal positivist Hans Kelsen: "What is a legal person
is for the law, including of course, the Constitution, to say, which
simply means that upon according legal personality to a thing
the law affords it the right and privileges of a legal person....
The point is that it is a policy determination whether legal personality should attach and not a question of biological or
'natural' correspondence."[21]

*Roe v. Wade* is the most blatant example of legal positivism.
The court declared that the fetus is not a person and therefore
has no rights. The Supreme Court chose the rationale used by
the Nazis: that an innocent human being can be declared a non-
person and deprived of life if "its" existence is merely incon-
venient or deemed *unfit* by those presently holding judicial or
legislative power.

The final absurdity is this: while the rights of the person
are denied, inanimate objects and animals are given rights. The
late Justice Douglas declared in *Sierra Club v. Morton* (1972), "The
voice of the inanimate object . . .should not be stilted."[22]

Philosopher Peter Singer insists that the pig has more consciousness and, therefore, more rights than fetuses or sick people.[23]

One is politically suspect today if one merely discusses the natural law in the public square. For instance, during the 1991 Senate Confirmation Hearings on Clarence Thomas, Professor Laurence Tribe of Harvard wrote in a *New York Times* op-ed piece that Thomas' "adherence to natural law as a judicial philosophy could take the [U.S. Supreme] Court in an even more troubling direction." Because Judge Thomas made references to the concept in speeches and articles, legal realists questioned his fitness to sit on the court. "Before the Senate decides whether to confirm Judge Thomas," Tribe concluded, "they should explore the implications of his views about natural law as the lodestar of constitutional interpretation."[24]

Senator Joseph Biden, then Chairman of the Judiciary Committee, stated that before the senate could decide whether Thomas was qualified sit on the Court, it would be necessary to find out if the nominee supported the "good" theory of natural law. The bad theory according to Biden is "a code of behavior," which suggests that natural law dictates morality to us. A good theory, presumably, leaves morality to individual choice.[25]

If the legal positivists are not checked, there will be no limit on their power to strip man and his family of their dignity and rights. A totalitarian state, not unlike Hitler's, can become a reality in our nation, and men will be the helpless prey of those in power. Returning to the natural law is the only means by which the power of the executive, legislative, and judicial branches of government can be curtailed and the inherent dignity of the human person upheld.

America needs the natural law if our democratic form of government is to survive. For it is the natural law that assures that government will be for the people and not the people for the government. It assures that the citizenry as well as those who control the daily operations of government will be subject to law. It assures that there is a distinction between good and evil. It assures that the murderer's taste for taking human life is treated differently than the physicians taste for preserving it.[26] The natural law assures that every man maintains his inalienable rights to life, liberty, and property.

This does not mean that the members of civil society can not possess positive legal rights. Natural law simply places a limit on the extent of experimentation in government. The distribution of power is largely irrelevant to the proponents of natural law so long as fundamental rights are protected. The natural law assures that the inherent dignity of every person is protected from the likes of a Hobbes or a Hitler or a Holmes. As Justice Felix Frankfurter pointed out: "The Constitution was intended, its very purpose was, to prevent experimentation with the fundamental rights of man."[27] The natural law is needed if the people want the civil government's ever changing rules and regulations to conform with the common good.

The only thing that will preserve our great democracy is that which gave it strength at its birth, a firm belief in the God-given dignity of each man and the fundamental principles connected with man's God-given nature. There are countless other things about whose values as means to attain the general welfare. Even the best minds may have doubts and differences. But the strength of our Union stems from our unity on first principles. Doubt our first principles, doubt everything, and we are doomed to division, distrust, discord, despair, and destruction. Doubt first principles, and we as a nation are doomed to that confusion and lack of concord which, as history has shown us, is the happy hunting ground of some form of Holmesian dictatorship.[28]

PART II

# ASSISTED SUICIDE
# AND EUTHANASIA

9

# EUTHANASIA THROUGH THE CENTURIES

*A society that believes in nothing can offer no argument even against death. A culture that has lost its faith in life cannot comprehend why it should be endured.*

Andrew Coyne

O N JANUARY 8, 1997 the United States Supreme Court heard arguments in support of a person's "right" to die. There can be no doubt that the Founding Fathers never contemplated any such right; theirs was after all the generation that had fought and died in defense of the *inalienable* right to life.

All across America challenges to this most basic principle have been raised: in ballot initiatives, in legislative debates, and in court decisions. And, to give the positivists their due, the controversy has raised many important philosophical questions. Are assisted suicide and euthanasia acts in defense of the quality of life, or are they simply licenses to kill? Is human life sacred no matter how diminished its quality? Is the individual's right to control over his body absolute? These and other questions will be addressed in this chapter.

Euthanasia (from the Greek *eu*, meaning well or good, and *thantos*, meaning death) was a term coined in the sixteenth century by proponents of "mercy killing" for people suffering lingering illnesses. *The New England Journal of Medicine* has defined the procedure as the deliberate administration of a lethal drug to hasten death in a suffering patient.[1] The reader should be aware that contrary to this definition many members of the medical and legal professions view assisted suicide and euthanasia as two different issues. For them assisted suicide means that a doctor provides medication or a drug-inducing apparatus (such as Jack Kevorkian has made famous) and the patient who then ends his own life. Euthanasia, on the other hand, is when the doctor ends the patient's life, usually with the injection of lethal drugs. In order to avoid confusion I will observe the distinction.

Opposition to euthanasia dates back to ancient times. The subject was central to the code of medical ethics devised by Hippocrates in 400 BC, and forms the basis of the Hippocratic Oath, which to this day is administered to medical candidates during graduation exercises. One part of the Oath states: "I will neither give a deadly drug to anybody if asked . . . nor will I make suggestions to this effect."[2]

Antipathy to euthanasia is also well established in the Judeo-Christian tradition. According to the Book of Genesis (9:6), "He who sheds man's blood, shall have his blood shed by man, for in the image of God man was made." And in Exodus (23:7), the law very specifically states that the "innocent and just person you shall not put to death." According to Dr. Immanuel Jakobovitis, former Chief Rabbi of England, Jewish Law states that while the killing of a person who suffered from a fatal injury (from other than natural causes) is not actionable as murder, the killer is morally guilty of a mortal offense.[3]

The Roman Catholic Church has always forbidden euthanasia and suicide. Responding to the Stoics, Epicureans, and Cynics, who looked upon suicide as a noble act, St. Augustine wrote in *The City of God*: "For it is clear that if no one has a private right to kill even a guilty man (and no law allows this), then certainly anyone who kills himself is a murderer, and is the more guilty in killing himself the more innocent he is of the charge on which he has condemned himself to death. . . . What we are saying, asserting, and establishing by all means at our command is this: that no one ought deliberately to bring about his own death by way of escaping from temporal troubles . . ."[4]

Numerous Church councils (first at Nimes, and then at Orleans, Braga, Auxerre, and Antisidor) forbade Christian burial to those who commit suicide, and this belief has been consistently reaffirmed by canon law.[5]

St. Thomas Aquinas postulated that the act of suicide violates the virtues of love, fortitude, temperance, hope, faith, prudence, and the legitimate prerogatives of the common good.[6] He also argued that it is an evil means to a good end when it is performed to escape suffering, shame, or an occasion of sin. In the *Summa Theologica* Aquinas wrote:

> Man is made master of himself through his free-will: wherefore he can lawfully dispose of himself as to those matters which pertain to this life as ruled by man's free-will. But the passage from this life to another and happier one is subject not to man's free-will but to the power of God. Hence it is not lawful for man to take his own life that he may pass to a happier life, nor that he may escape any unhappiness whatsoever of the present life, because the ulti-

mate and most fearsome evil of this life is death, as the Philosopher [Aristotle] states . . . Therefore, to bring death upon oneself in order to escape the other afflictions of this life, is to adopt a greater evil in order to avoid a lesser.[7]

In 1940, 1943, and 1948 Pope Pius XII condemned both compulsory and voluntary mercy killing for any and all reasons— particularly economic and racial reasons.[8] In 1957 the Church declared that there was a substantial difference between ordinary and extraordinary measures of continuing a sick person's life.[9] For example, the administration of drugs in large doses for the purpose of killing pain in terminally ill patients is permissible, even though the unintended consequence is the shortening of life. The Second Vatican Council affirmed the Church's traditional teaching: "Whatever is opposed to life itself such as any type of murder, genocide, abortion, euthanasia or willful self destruction . . . all these things and others of their like are infamies indeed."[10] On May 5, 1990 the Vatican issued a *Declaration on Euthanasia* which states:

1. No one can make an attempt on the life of an innocent person without opposing God's love for that person, without violating a fundamental right, and therefore without committing a crime of the utmost gravity.

2. Everyone has the duty to lead his or her life in accordance with God's plan. That life is entrusted to the individual as a good that must bear fruit already here on earth, but that finds its full perfection only in eternal life.

3. Intentionally causing one's own death . . . is therefore equally as wrong as murder, [and] such an action on the part of a person is to be considered as a rejection of god's sovereignty and loving plan. Furthermore, suicide is also often a refusal of love for self, the denial of the natural instinct to live, a flight from the duties of jus-

tice and charity owed to one's neighbor, to various communities or to the whole of society—although, as is generally recognized, at times there are psychological factors present that can diminish responsibility or even completely remove it. However, one must clearly distinguish suicide from that sacrifice of one's life whereby for a higher cause, such as God's glory, the salvation of souls or the service of one's brethren, a person offers his or her own life and puts it in danger.[11]

The Vatican's statement concluded with these comments: "Furthermore, no one is permitted to ask for this act of killing, either for himself or herself or for another person entrusted to his or her care, nor can he or she consent to it, either explicitly or implicitly. Nor can any authority legitimately recommend or permit such an action. For it is a question of the violation of the divine law, an offense against the dignity of the human person, a crime against life, and an attack on humanity."[12]

Analyzing the Church's position, Timothy O'Donnell, in his work *Medicine and Christian Morality,* wrote:

> The Catholic Church has consistently taught that suicide is a totally indefensible and gravely sinful injection of disordered self-referenced determination into the providential plan of God's love. It is seen as the ultimate violation of the divine prerogative of the Author of life. We might add that the divinely endowed rights which the Founding Fathers of American democracy held to be self-evident... carry with them certain corresponding fundamental responsibilities. The most profound of these responsibilities is drastically and definitively abandoned in the act of self-destruction, and, indeed, in this context euthanasia and suicide are very much of a piece, and present the same theological distortion of the right order.[13]

The Anglican Church also distinguishes between ordinary and extraordinary care in decisions about life and death:

> "Ordinary" in this context does not mean what a medical man would regard as "normal" treatment: It means whatever a patient can obtain and undergo without thereby imposing an excessive burden on himself or others. Thus, extraordinary treatment has been defined as "whatever here and now is very costly or very unusual or very painful or very difficult or very dangerous." . . . The doctor has neither right nor duty to insist on extraordinary treatment against the patient's will, nor is he bound to apply such treatment in cases where the patient cannot be consulted; and the patient's family is in much the same position.[14]

Although the concept of euthanasia was discussed for centuries in the western world's most fashionable salons, it was not until the late nineteenth century when the Social Darwinists dominated the intellectual scene that it was taken seriously by the medical and legal professions. The secularist philosophy known as Social Darwinism holds that the progress of a society is rooted in the laws of physics and biology and that advancement in society is achieved by those most fit to survive. Historian Richard Hofstadter described the Social-Darwinist position this way:

> Domination by the fittest is of the greatest benefit to society as a whole. . . . To progress, a social system must retain competition between the directors of labor, the contest for industrial domination. No matter what happens to society, the domination of the fittest great men—capitalistic competition—must be ensured. Such men are the true producers. The fundamental condition of social progress is that these leaders be obeyed by the masses. In politics, as in industry, the forms of democracy are hollow; for while executive agencies are designed to execute the will of the many,

the opinions of the many are formed by the few, who manipulate them.[15]

America's leading apostle of Social Darwinism, William Graham Sumner of Yale, declared: "Let it be understood that we cannot go outside of this alternative: liberty, inequality, survival of the fittest; not liberty, equality, survival of the unfittest. The former carries society forward and favors all its best members; the latter carries society downwards and favors all its worst members."[16]

For the Social Darwinist, people who are judged an economic or medical burden on society may be slated for elimination.[17] American supporters of this view founded numerous organizations, including the Eugenics Record Office and the Cold Spring Harbor Eugenics Laboratory (funded by the Rockefellers, Harrimans, and Carnegies), and promoted euthanasia legislation throughout the nation.[18]

One euthanasia bill was introduced in the 1906 session of the Ohio State Legislature. It stated that if a person was "seriously ill or injured and had the mental competence to ask for death and three physicians agreed that the patient was in a terminal state, then it is permissible to end the life."[19] The bill was overwhelmingly rejected 78 to 22.

The Euthanasia Society was founded in 1938. Its charter contained this preamble: " . . . with adequate safeguards the choice of immediate death rather than prolonged agony should be available to the dying."[20]

It is important to emphasize that within this century the arguments for assisted suicide and euthanasia have always gone beyond simply complying with the wishes and concerns of the patient. Distinguished organizations and members of the medical and legal professions have also claimed that the physician, not just the patient, should be empowered to determine if the patient's life should be ended. Dr. Robert Proctor, in his work *Racial Hygiene*, points out that while in earlier times "advocates of euthanasia defended the right to choose the time and manner of one's death or to end one's life with a minimum of pain and suffering . . . [i]n the twentieth century . . . euthanasia has been recommended as means of cutting costs or ridding society of 'useless eaters'."[21]

In the 1938 edition of the *Journal of the American Institute of Homeopathy*, Dr. W. A. Gould proposed euthanasia as a means of relieving economic hardship and suggested that this was no more than the elimination of the unfit.[22] Dr. Foster Kennedy of Cornell Medical School wrote in the *Journal of the American Psychiatric Association* (1942) that assisted suicide should be applied to "those helpless ones who should never have been born—nature's mistakes."[23] Dr. Kennedy, who accepted an honorary doctorate from a Nazi university in 1936, resigned from the Euthanasia Society of the United States in 1939 "because he criticized its policy of favoring 'voluntary' euthanasia for people who had at one time been well, but later became ill. In contrast, he favored systematic extermination."[24] The inventor of the iron lung, the French-American Nobel Prize winner Alexis Carrel, wrote in his book *Man the Unknown* (1935) that the incurably sick and the insane should be "*humanely* and economically disposed of in small euthanasia institutions supplied with proper gases."[25] [Italics added.]

In the early twentieth century, medical groups throughout the United States and Europe proposed euthanasia legislation that abrogated the rights of the patient in favor of the doctor or the state. Nowhere was this more true than in Nazi Germany.

# 10

# THE GERMAN EUTHANASIA EXPERIENCE:
# THE FINAL SOLUTION

*Whatever proportions these crimes finally assumed, it became evident to all who investigated them that they had started from small beginnings. The beginnings at first were merely a shift in emphasis in the basic attitude of the physicians. It started with the acceptance of the attitude, basic in the euthanasia movement, that there is such a thing as life not worthy to be lived. This attitude in its early stages concerned itself merely with the severely and chronically sick. Gradually the sphere of those to be included in this category was enlarged to encompass the socially unproductive, the ideologically unwanted, the racially unwanted and finally all non-Germans.*

Leo Alexander, M.D., American
Medical Consultant at Nuremberg

THE GERMAN MEDICAL ESTABLISHMENT, influenced by Social-Darwinist theories, aggressively promulgated assisted suicide practices. For them it was imperative for the state to stop the "degeneration" of mankind. In his 1895 book *The Right to Death,* Dr. Adolph Jost argued that the individual's fate belonged to the state. While he spoke of compassion and relief of the suffering of the incurably ill, his focus was mainly on the health of the *Volk* (the people) and of the state.[1] "The state," Jost declared, "must own death—must kill—in order to keep the social organism alive and healthy."[2] At the same time, Dr. Alfred Ploetz publicly discouraged medical care for the weak and incurably ill and founded (1905) the Society for Racial Hygiene. During the 1920s and 1930s branches were formed throughout the Weimar Republic and membership, limited to "Aryan" doctors, soared.[3] Dr. Ploetz was recognized in the 1930s as the creator of the biological foundations for the Nazi racial state.

*Release and Destruction of Lives Not Worth Living* (1920) by medical professor Alfred Hoche and law professor Rudolf Binding argued for "allowable killing" of the terminally ill.[4] "The right to live," they claimed, "must be earned and justified not dogmatically assumed."[5] They insisted that assisted suicide was a *human right* and that killing the physically unfit was purely "*healing* treatment" or "*healing* work."[6] [Italics added.]: "[They] used the argument that the terminally ill deserved the right to a relatively painless death to justify the murder of those considered inferior. Binding and all subsequent proponents of his argument consciously confused the discussion by pointing to the suicide rights of terminal cancer patients facing a certain and painful death when in reality they wanted to "destroy" the "unworthy life" of healthy but "degenerate" individuals."[7]

The German medical profession embraced the National Socialists—for them, Nazism was "applied biology."[8] Dr. Herman

Berger commented in 1934: "The National Socialists Physicians' League proved its political reliability to the Nazi cause long before the Nazi seizure of power, and with an enthusiasm . . . and an energy unlike that of any other professional group."[9] More than forty-five percent of German physicians were members of the Nazi Party.

Germany's most prestigious medical journal *Deutsches Arztoblatt* stated (1933) that "never before has the German medical community stood before such important tasks as that which National Socialists ideas envision for it."[10] Indeed, German medical scholarship provided the ideas and techniques which led to and justified unparalleled slaughter.[11] It led to policies, implemented by German Euthanasia Program Director Victor Brackhead, whose basic assumption was that the needle belongs in the hands of the doctors.[12] After all, as philosopher Friedrich Nietzsche said: "The sick person is a parasite of society."

Hitler himself was certainly a strong supporter of euthanasia. In *Mein Kampf* (1923) he wrote: "The right of personal freedom recedes before the duty to preserve the race. There must be no half measures. It is a half measure to let incurably sick people steadily contaminate the remaining healthy ones. This is in keeping with the humanitarianism which, to avoid hurting one individual, lets a hundred others perish. If necessary, the incurably sick will be pitilessly segregated . . . [a] barbaric measure for the unfortunate who is struck by it, but a blessing for his fellow men and posterity."[13]

As early as 1935 Hitler told Dr. Gerhard Wagner, head of the Nationalist Socialist Physician's League, that "large scale euthanasia would have to wait until wartime because it would be easier to administer."[14] In October, 1939 Hitler signed the following memo: "Reichsleiter Bouhler and Dr. Brandt are charged with the responsibility for expanding the authority of physicians who are to be designated by name, to the end that patients who, in the best available human judgment after critical evaluation of their condition are considered incurable, can be granted *Gnadentod* [mercy death]."[15] Hitler was very clever: to avoid any direct responsibility for the program, he did not order doctors to employ involuntary assisted-suicide measures; he simply empowered them to do so. Nevertheless, the German Medical Society enthusiastically complied. When Hitler gave German doctors a license to deal with handicapped people—free from the restraints of law or ethics—and when he allowed the doctors to proceed in secret away from the cleansing force of public awareness, the members of the medical establishment were at liberty to act upon their most base and primitive feelings concerning the disadvantaged.[16]

Six euthanasia centers were opened and given the euphemistic designation, Charitable Foundations for Institutional Care. As the name of the centers suggests, killing was rationalized as a *compassionate* act. Although the justifications of euthanasia almost always included the word *mercy*, the real motivation for the killings was economics: "It must be made clear to anyone suffering from an incurable disease that the useless dissipation of costly medications drawn from the public store cannot be justified. Parents who have seen the difficult life of a crippled or feeble-minded child must be convinced that, though they may have a moral obligation to care for the unfortunate creature, the broader public should not be obligated . . . to assume the enormous costs that long-term institutionalization might entail."[17]

It was revealed at the Nuremberg Trials that this initial program, confined to Germany, was responsible for the deaths of at least 70,000 adults and 5,000 children, although other estimates are as high as 400,000.[18] The program was terminated by Hitler's orders in 1941 because there was genuine opposition throughout the Reich.[19] August von Galen, the Catholic Bishop of Munster had the greatest public impact when he denounced the euthanasia policies from the pulpit in August 1941. The bishop said:

> If you establish and apply the principle that you can "kill" unproductive human beings, then woe betide us all when we become old and frail! If one is allowed to kill unproductive people, then woe betide the invalids who have used up, sacrificed and lost their health and strength in the productive process. . . . Poor people, sick people, unproductive people, so what? Have they somehow forfeited the right to live? Do you, do I have the right to live only as long as we are productive? . . . Nobody would be safe anymore. Who could trust his physician? It is inconceivable what depraved conduct, what suspicion would enter family life if this terrible doctrine is tolerated, adopted, carried out.[20]

There were also public protests—a rarity in Nazi Germany.

The solution was to move the programs to the conquered eastern nations where they became known as the Final Solution. Since Jews, Gypsies, and Slavs were considered "diseased" races, involuntary assisted suicide was the prescribed "medical means" to eliminate them.

It was the verdict of the Nuremberg Trials that the German euthanasia program was the direct precursor of Nazi genocide.[21] The judges did not accept the defense argument that doctors and nurses (at the Haldamar Euthanasia Hospital for instance) who were carrying out the laws of the land were thus shielded from punishment.[22] Nor were the judges swayed by one euthanasia doctor's assertion that "from a medical standpoint it is a humane motive to shorten the lives of children not fit to live."[23] The accused were found guilty of violating the natural law and were punished.

Chief American counsel Robert H. Jackson viewed the progression of German euthanasia this way:

> A freedom-loving people will find in the records of the war crimes trials instruction as to the roads which lead to such a regime and the subtle first steps that must be avoided. Even the Nazis probably would have surprised themselves, and certainly they would have shocked many German people, had they proposed as a single step to establish the kind of extermination institution that the evidence shows the Haldamar Hospital became. But the end was not thus reached; it was achieved in easy stages.
>
> To begin with, it involved only the incurably sick, insane and mentally deficient patients of the institution. It was easy to see that they were a substantial burden to society, and life was probably of little comfort to them. It is not difficult to see how, religious scruples apart, a policy of easing such persons out of the world by a completely painless method could appeal to a hard-pressed and unsentimental people. But "euthanasia" taught the art of killing and accustomed those who directed and those who administered the death injections to the taking of human life. Once any scruples and inhibitions about killing were overcome and the custom was established, there followed naturally an indifference as to what lives were taken. Perhaps also those who become involved in any killings are not to be in a good position to decline further requests. If one is convinced that a person should be put out of the way because, from no fault of his own, he has ceased to be a social asset, it is not hard to satisfy the conscience that those who are willful enemies of the prevailing social order have no better right to exist. And so Haldamar drifted from a hospital to a human slaughterhouse.[24]

With the revelation of the horrendous consequences of the German euthanasia programs, the American assisted-suicide movement went underground. The Cold Spring Harbor Eugenics Laboratory, for instance, dropped "Eugenics" from its title, apparently in an attempt to maintain respectability. By the 1960s and 1970s, however, the movement was making a comeback, prompting this comment from journalist-philosopher Malcolm Muggeridge: "For the Guinness Book of Records, you can submit this: that it takes about 30 years in our humane society to transform a war crime into an act of compassion."[25]

Joseph Fletcher, creator of "situation ethics" and a board member of the Euthanasia Education Council, wrote in the *American Journal of Nursing* (1973): "If we are morally obligated to put an end to a pregnancy when an amniocentesis reveals a terribly defective fetus, we are equally obligated to put an end to a patient's hopeless misery when a brain scan shows that a patient with cancer has brain metastases."[26] To appear more compassionate and less strident the Euthanasia Education Council changed their name to Concern for Dying, Inc.; and the Euthanasia Society of America became the Society for the Right to Die, Inc. Other organizations appeared including Choice in Dying, Inc. and Americans Against Human Suffering. *Euthanasia News* reported in February 1975: "Experience has shown that legislators and lawyers have expressed gratitude for receiving material on 'death with *dignity*' but object to receiving it from an organization with the word 'euthanasia' in its title. Secondly, it was felt that 'The Right to Die' more closely approximates what the society stands for."[27]

Although the progress of the American euthanasia movement has been notable, it can only marvel at the process by which suicide and even murder have become fundamental human rights in the Netherlands.

# 11

## THE DUTCH EUTHANASIA EXPERIENCE: ANYTHING GOES

*We Dutch pride ourselves on our history. We see ourselves as having been good in the past, therefore, we believe that we will always be good. It is an arrogance of goodness. Thus, even through there are striking resemblances to our euthanasia practices and those the Nazis sought to impose upon us, we assure ourselves: We resisted the Nazis. We are sophisticated, humane. We can't be doing wrong! To admit we are wrong on euthanasia would be to say that we are not the compassionate, sophisticated, enlightened people we think we are. It is very hard in the Dutch character to do that.*

W. C. M. Klijn
Dutch Professor of Medical Ethics

THE GOVERNMENT, medical, and legal professions in the Netherlands take the concept of "tolerance" very seriously. Viewing themselves as free of traditional restrictions they tolerate drugs, tolerate prostitution, tolerate pornography, and tolerate euthanasia. The eminent Dutch historian Johan Huizinga, author of *The Spirit of the Netherlands*, made these comments concerning his homeland's attitudes: "Tolerance is a virtue that can become a vice. Respect for the rights and opinions of others too often leads to respect for their wrongs. . . . The belief that what is evil becomes good if only enough people want it is one of the most terrifying observations of the age."[1] The tolerance of euthanasia is indeed frightening, and the progress of its acceptance in the Netherlands has followed a predictable course: first there was assisted suicide; then there was euthanasia of the terminally ill and next of the psychologically unhappy; and finally there was involuntary euthanasia, or "termination of the patient without explicit request."[2]

For over one hundred years the Dutch criminal code had forbidden euthanasia for competent terminally ill patients. The old rules, enacted in 1886 and still on the books to this day, state: "He who robs another of life at his express and serious wish is punished with a prison sentence of at most 12 years.... He who deliberately incites another to suicide, assists him therein or provides him with means, is punished if the suicide follows, with a prison sentence of at most three years."[3] A memo attached to the code further explains its intent: "The assent cannot abrogate the criminalization of taking a life, but [can] give it a wholly different character. The law as it were no longer punishes the attack against the life of a particular person, but the violation of the respect which is due human life in general, regardless of the motive of the perpetrator. Crime against life remains, the attack on the person is abrogated."[4] In recent decades, however, Dutch court decisions, government-sponsored medical commissions, and the political and legal professions have created so many loop-holes that de facto euthanasia now exists in the Netherlands.

In the 1970s Dutch courts permitted physician-assisted suicide for terminally ill competent patients. In 1973 the Royal Dutch Medical Association, while affirming its continued support for the penal articles that prohibited euthanasia, added this comment: ". . . combating pain and discontinuing futile treatment could be justified, even if the patient died as the result of the act or omission." [5]

In a 1981 decision the Rotterdam Court listed nine criteria for justified euthanasia:

1. The patient must be suffering unbearably.
2. The patient must be conscious when he expresses the desire to die.
3. The request for euthanasia must be voluntary.
4. The patient must have been given alternatives with time to consider them.
5. There must be viable solutions for the patient.
6. The death must not inflict unnecessary suffering on others.
7. The decision must involve more than one person.
8. Only a physician may perform the euthanasia.
9. The physician must exercise great care in making the decision. [6]

Strong as they are, these criteria represent just a small opening in the door that was formerly closed on euthanasia. Throughout the 1980s and 1990s the door swung ever wider.

In 1984 the Netherlands Supreme Court sanctioned physician-assisted suicide for chronically-ill and elderly patients who were not in danger of imminent death. The Royal Dutch Medical Association, working with the nation's nursing associations, issued these updated "Guidelines for Euthanasia":

1. The patient must be a mentally competent adult.
2. The patient must request euthanasia voluntarily, consistently, and repeatedly over a reasonable time, and the request must be documented.
3. The patient must be suffering intolerably, with no pros pect of relief, although the disease need not be terminal.
4. The doctor must consult with another physician not in volved in the case.[7]

Once enacted into law these guidelines encouraged the judiciary to further expand the pool of potential candidates for euthanasia. In 1989 the Dutch Supreme Court chose not to prosecute a doctor who gave a lethal injection to a newborn with Down's Syndrome. The court reasoned that "since the child would have experienced very serious suffering after surgery, it was not likely that the physician would be convicted if his case went to court, and therefore, his objection to prosecution was justified."[8]

By the early 1990s the judiciary was permitting assisted suicide for psychiatric patients who were physically fit. In one case a psychiatrist was judged not guilty of assisting in a suicide of a physically healthy individual because the court concluded that the patient, although suffering from a mental illness, was competent and completely free to make the choice to die. The court deemed that it would be *discriminatory* to permit assisted suicide only in cases of people who suffer physically. Psychological pain or even unhappiness cannot be excluded as valid reasons for suicide.[9]

Dutch courts have often ruled that when a doctor's conscience is in conflict with the law he is permitted to prescribe euthanasia to relive suffering. This is justified as an example of a *force majeure*—an unforseen course of events that abrogates the usual legal necessities.[10]

To determine if laws and guidelines pertaining to euthanasia were being adhered to, the government-sponsored Remmeling Commission was established. Its findings, published in 1991, were startling:

* Of the 130,000 annual deaths, 49,000 involve a medical decision at the end of life, and in 95 percent of these cases there is a withholding or discontinuing of life support or the use of painkilling medications to hasten death.

* 50 percent of the 49,000 termination decisions were made without the patient's permission.

* 2,700 cases were direct euthanasia cases.

* 1,000 of the 2,700 direct euthanasia cases were performed without any request from the patient.

* 27 percent of physicians terminated lives with patient approval.

* 40 percent of the doctors did not keep records of euthanasia cases.

* 71 percent of the physicians did not list euthanasia on the death certificate.[11]

Reacting to these appalling figures the commission concluded that they were not "morally troublesome because the suffering of the patients involved had become unbearable and they would usually have died soon anyhow."[12]

Dr. Herbert Hendin, Executive Director of the American Suicide Foundation, commented that the commissioners were "concerned with protecting [euthanasia] as a right; as a result they seemed more supportive of doctors who practice it than protective of patients."[13]

In the Netherlands today courts permit a physician to unilaterally intervene in cases where family members cannot make a decision. A doctor need only ask himself "if he would accept life if he were in the patient's position," and if he *knew* (although not necessarily had consulted) another doctor who would agree that under the given circumstances the patient's life is not worth preserving or is a "limited life."[14]

Dr. Richard Fenigsen, a Dutch cardiologist, described what is now a typical experience for physicians in his homeland:

> Mrs. P. was a seventy-two-year-old widow who, after a bad myocardial infarction, was left with a grossly enlarged heart and congestive heart failure. She was treated . . . and for a whole year had almost no symptoms at rest. True, she needed help with cleaning the house, and her only exercise was walking a few blocks. [Then] . . . her breathlessness recurred . . . Another time she complained of dizziness, which turned out to be due to a fall in blood pressure in an upright posture; she was taught the necessary precautions. Mrs. P. was an extremely nice, mild-tempered lady who never showed any impatience and complied with the doctor's every order and advice. Barring some clot or a sudden disturbance in heart rhythm (both of which could of course occur), she might have remained for years in the condition she was in. When she once failed to appear at the outpatient clinic, I was very much worried. Responding to my inquiry, her family physician, Dr. De K., paid me a visit. He had a talk with Mrs. P., he said, and explained the situation to her: This wasn't going to be any better and living such a limited life, with all those pills, made no sense at all. Mrs. P. accepted everything he said. He stopped her pills, and three days later she died. My only answer was to nod; I couldn't emit a sound. I was overcome by deep sorrow. It returns every time I think of Mrs. P.[15]

Dutch Doctors are virtually unaccountable. Ethicist Joe Welie of the University of Nigmegun conceded that Dutch physicians have been elevated to a "superior moral status making their judgments on life and death always just." [16]

This godlike power even touches the lives of children. Researching his essay, *Physician-Assisted Death in the Netherlands* (1995), Fenigsen discovered that Dutch tolerance does not extend to disabled children. Here are descriptions of several cases he uncovered:

- A girl born prematurely, in the thirty-second week, recovered from an infection, but there was a suspicion of intracranial bleeding. This was followed by accumulation of intracranial fluid. The parents refused to allow the insertion of a drainage tube or shunt. On the thirtieth day after birth the child was killed by the pediatrician with injections of a morphine-like drug and potassium chloride. . . .

- Danny had spina bifida and hydrocephalus but was in fair general condition. No drainage tube to relieve the hydrocephalus was inserted. Once Danny seemed to have some abdominal pain, and another time he apparently felt not quite well for two consecutive days. This prompted the parents to ask for euthanasia. With this purpose the child was admitted to Rainier de Graaf Hospital in Delft. One of the nurses opposed the decision, and on the next day she and her husband offered to adopt the child. The offer was rejected. On August 19, 1990, Danny, then aged three and one-half months, was killed with drugs administered by intravenous drip. The nurse was reprimanded because by involving her husband in the adoption offer she violated professional confidentiality. . . .

- This six-year-old boy's intelligence seemed below average. His upbringing presented some problems. He lived with his parents and attended a school for children requiring special care. Then juvenile diabetes mellitus was discovered. Patients with this type of diabetes must receive injections of insulin; otherwise they develop severe disturbances in metabolism (ketoacidosis), become comatose, and die. The family physician did not ask the parents for permission to start the insulin treatment. Instead, he asked them whether their son should be treated. The parents chose not to treat the child, and the boy died.[17]

Doctors are no longer neutral observers. They can offer euthanasia as one or the only option; they do not have to distinguish between competent and non-competent patients; they alone decide when a person's remaining time is not valuable; they alone decide if a patient's life is too costly to maintain or a nuisance or a burden on the family. Advocates of euthanasia see no difference between a doctor's withdrawing futile treatment and his becoming a more direct participant in inducing death.[18]

Some may find this "Orwellian," especially those who dissent from the new right-to-die orthodoxy. Dutch doctors who oppose euthanasia risk being brought up on disciplinary charges if they refuse to make proper referrals. Traveling pools of doctors (known as the "Angels of Death") are permitted to go out and employ euthanasia when a local physician or family refuses the "treatment." Euthanasia is sometimes performed without the knowledge of treating physicians, and some non-medical volunteers are allowed to give lethal injections.[19] "Euthanasia advocates," writes Dr. Herbert Hendin, "dismiss medical and community opposition, attributing it to religious conservatives and to some who lost families in Nazi concentration camps; both groups are said by advocates to be unable to be objective about euthanasia."[20] Dr. Isaac van der Sluis, a non-religious opponent of euthanasia, sees "advocates as portraying themselves as on the 'side of angels' defending a 'beautiful liberal ideal' while behaving as though in a war where any criticism of Dutch euthanasia policies served the cause of the enemy."[21]

# 12

# THE AMERICAN EUTHANASIA EXPERIENCE:
# THE SLIPPERY SLOPE

*Despite the technological revolution, we physicians
must continue to honor a tradition that has preserved
for thousands of years; the necessity to preserve the
best possible life for the longest possible time. When
one backs away in any sense from the utter sanctity of maintaining human life, the slope becomes
very slippery indeed.*

George Lundberg, M.D.,
*Journal of the American Medical Association*

CAN THESE Dutch horrors happen in the United States?
Sad to say, we are already well down the road. Consider
these facts:

- In a 1971 issue of the *Humanist* Dr. Walter C. Alvarez
  wrote: "Some day a law will be passed saying that questions of when to pull out the tubes must be left in the

hands of the patient's physician or physicians. Such a law might save the country a few billion dollars a year."[1]

◆ The *Florida Sun* reported on January 11, 1973 that Dr. Walter Sackett projected that "$5 billion could be saved in the next half century if the state of Florida's mongoloids [individuals with Down's Sydrome] were permitted merely to succumb to pneumonia."[2]

◆ In a 1977 Department of Health Education and Welfare memo outlining health-care cost reductions, agency administrator Robert Derzon explained that mandatory living-will laws would make it easier to terminate life support and estimated that this would save the Medicare system $1.2 billion annually.[3]

◆ Nobel Laureate Francis Crick (co-discoverer of DNA) told the Pacific News Service in January 1978 that "no newborn infant should be declared human until it has passed certain tests regarding its genetic endowment and that if it fails these tests it forfeits the right to life."[4]

◆ Two doctors associated with Yale University told the *Milwaukee Journal* in 1971 that staff physicians in the Yale-New Haven Hospital in Connecticut "have helped parents give defective infants lethal drugs overdoses."[5]

◆ An anonymous essay entitled "It's Over Debbie" in the *Journal of the American Medical Association* (January 1988) described how a physician killed a 20 year old cancer patient with a fatal dose of morphine after hearing the patient mumble "Let's get this over with."[6]

◆ The February 22, 1988 edition of *U.S. News and World Report* quotes ethicist Daniel Callahan, director of the Hastings Center: "What do we owe each other as we grow old? I think we would be justified in saying that beyond a certain age we will simply not provide expensive, life extending care."[7]

• In 1988 the Hemlock Society's Executive Director Derick Humphry burned the responses of a *Journal of the American Medical Association* poll to avoid any official inquiries into the 78 doctors who admitted to having committed euthanasia or, as Mr. Humphry called it, *"compassionate crimes."*[8]

• In the *New England Journal of Medicine* (1991) Dr. Timothy Quill described how he provided lethal drugs to "Diane," a leukemia patient, who preferred to die instead of taking long-term therapy.[9]

• In March, 1996 the American Medical Association's Council on Ethics and Judicial Affairs published an advisory statement that sanctioned the termination of non-terminally ill coma patients on life-sustaining technological equipment. Food supplied through tubes was redefined as such a technology, and the guidelines permit a doctor to withdraw food, thus allowing a patient to die from starvation and dehydration.[10]

• A 1996 study released in the *New England Journal of Medicine* revealed that 19 percent of intensive care nurses who responded to a poll admitted that they chose to hasten the death of terminally ill people without notifying the patient, family or presiding physician. Five nurses admitted helping patients die outside the hospital and 40 percent said they wanted to commit euthanasia but were afraid they would get caught.[11]

• A 1996 Washington State Study indicated that 24 percent of the doctors who responded to the survey admitted complying with euthanasia requests.

• Speaking at the August 1996 meeting of the Albany Humanist Society, H. William Batt, president of the New York State Hemlock Society, said: *"Dignity* is every person's right in a civilized society. . . . When we reach the point that we no longer have the capacity to choose to be, should

we not have the right—if not the responsibility—to excuse ourselves from society and the world?"

• A July 15, 1996 *New York Times* headline read: Many turning
to Internet for aid with suicide: Computers open new av-
enues for people who want to die.[12]

• Oregon voters approved the 1997 "Death with Dignity
Act" ballot initiative that permits doctors, with a second
opinion, to write prescriptions for drugs that they know
patients will use to commit suicide.[13]

And then there is Dr. Kevorkian . . .

LABELED BY HIS medical school classmates as "Dr. Death" because his hobby was to photograph patients' retina blood ves-
sels at the moment of death, Dr. Jack Kevorkian's practice has
certainly raised public awareness of the debate on assisted suicide.

Kevorkian made his public debut in 1990, when, after a brief
meeting with Janet Adkins (who was diagnosed to be in the early
stages of Alzheimer's disease), he agreed to aid her in committing suicide. Although reprimanded, he was not prosecuted because Michigan, unlike thirty-five other states, did not have a
statue forbidding assisted suicide.

The reprimand did not stop him. He participated in two
assisted suicides in 1991, five in 1992, eleven in 1993. The
Michigan State legislature reacted by passing a restrictive law and
Kevorkian was indicted, and he has been tried and acquitted three
times since 1994.

The prosecutors failed to convict Kevorkian—under the terms
of the new law or under common law—because they failed to
prove that he actually intended to help people kill themselves,
and because Kevorkian successfully convinced the jury that his

goal was "to relieve intolerable pain and suffering . . . to remedy their [i.e., the patients'] anguish, their torture. He testified that the patient "always has the absolute autonomy in these cases. . . . This seems in the patient's mind the only alternative. I have an *ethical* obligation to *honor* that autonomous decision. . . . I don't think I'm doing anything wrong. I know I'm not. . . . Is it my intention for them to die when there are tears going down my cheek? . . . My aim is not to end human life." [Italics added.] Kevorkian admitted that he brought a gas mask into the room, and even that he helped the patient "put it in place . . ." But, he concluded, it was the patient who "started the flow of carbon monoxide."[14]

Kevorkian apparently contemplated the subject of assisted suicide for many years. Early in his career he wrote numerous papers urging that criminals waiting on death row be "used as human guinea pigs." He noted that human experiments on criminals would save the lives of *innocent* animals killed in the name of science. In a 1991 work entitled *Prescription Medicide: The Goodness of a Planned Death*, Kevorkian introduces the term "obitiatry," the practice of experimentation on living humans while they are under anesthesia and prior to the imposition of medicide. Dr. Death states that his "ultimate aim . . . is not simply to help suffering or doomed persons kill themselves—that is merely the first step, an early distasteful professional obligation (now called medicide) that nobody in his or her right mind could savor. . . . [What] I find most satisfying is the prospect of making possible the performance of invaluable experiments or other beneficial medical acts under conditions that this first unpleasant step can help establish—in a word, obitiatry . . ."[15]

Kevorkian also calls for the creation of boards that would certify obitiatrists trained in medicide. If implemented his plans would establish zones within a given state for obitiatry headquarters and death clinics, plans eerily reminiscent of the Nazis' Charitable Foundations for Institutional Care.

Dr. Death's killing spree continues. As of July 1997 he had assisted in over forty suicides. But what is frightening is that autopsies of his victims reveal that most were not terminally ill. Consider the following examples:

- Suicide #1, Janet Adkins: she suffered from the early stages of Alzheimer's and had played tennis several days before her June 4 suicide.

- Suicide #3, Marjorie Wantz: she had a history of suicide attempts and complained of pelvic pain, but the autopsy did not indicate the presence of a terminal disease.

- Suicide #29, Ruth Neuman: the coroner's office stated, "what ever they claim, she was not terminally ill."

- Suicide #33, Rebecca Badger: the autopsy revealed that, contrary to what she claimed, she did not suffer from multiple sclerosis.

- Suicide #35, Judith Curren: she was overweight, tired, depressed, and her family had a history of domestic violence, but she did not have a terminal illness.[16]

Dr. D. J. Dragovic, Oakland County Michigan's medical examiner, whose office performed autopsies on twenty-seven of Kevorkian's first thirty-five cases, said that "at least half had serious questions about being terminal," and only four or five he said, "had just weeks to live. . . . There were a lot of people physically incapacitated that could have lived for many months to many years."[17]

Reacting to Kevorkian's activities, assisted suicide advocate Timothy Quill wrote in a *New York Times* op-ed piece (August 29, 1996): "Jack Kevorkian is on a rampage, and many who believe in doctor-assisted suicide are as horrified by what he is doing as are those on the other side of the debate."[18] The AMA declared that "no civilized society should condone assisted suicide as it is practiced by Jack Kevorkian", and called him "a reckless instrument of death and a great threat to the public."[19]

Such comments have not deterred Kevorkian. In a speech before the National Press Club in July 1996 he stated: "Pass any law you want, I don't care. . . . Had Christ died in my van with people around him who loved him [it] would have been far more dignified."[20]

# 13

## COURTING DEATH:
## RECENT LEGAL DECISIONS
## ON ASSISTED SUICIDE

*We rely on the sane people of the world to preserve
it from barbarism, madness, and destruction. Now
it begins to dawn on us that it is precisely the sane
ones who are the most dangerous.*

Thomas Merton

BY THE LATE 1980s American courts had entered the fray.
Here is a recap of some of the more notable recent cases
in the area of assisted suicide:

### CRUZAN VS. DIRECTOR, MISSOURI DEPARTMENT OF HEALTH

As a result of a January 1983 car accident, Nancy Cruzan suffered from profound cognitive disability. Although she required
intensive-care hospitalization, she was not on a respirator and
was not considered terminally ill. Nancy was able to swallow

food and there were indications that she could see and hear. To make it easier to care for Nancy, hospital attendants placed her on a feeding tube.

In 1988 Nancy's parents filed suit against the hospital after administrators refused to carry out the family's request that the food tubes be removed so their daughter might be permitted to die. Jasper County Circuit Court judge Charles Teel ordered the hospital to comply with the family's wishes, but the Missouri Supreme Court quashed Teel's order stating: "This is not a case in which we are asked to let someone die . . . This is a case in which we are asked to allow the medical profession *to make* Nancy die by starvation and dehydration."[1]

In 1990 the United States Supreme Court ruled that since there was not "clear and convincing evidence" that Nancy would have chosen death, the hospital would have to continue caring for her. However, although Chief Justice Rehnquist, speaking for the majority, distinguished between the act of assisted suicide and the termination of life-support systems, the court also accepted the notion that artificially supplied food and liquids are a form of medical treatment.[2] (Will a baby's bottle or a straw be the next feeding conduits to be declared medical treatments?)

This decision did not stop the Cruzan family. Suddenly they found two people (both had worked with Nancy) who were willing to testify in a Missouri court that they heard Nancy make comments, years before, that she never wanted to live in a coma. That was enough for Judge Teel, who ruled that the hospital must comply with the Cruzan family's request. Twelve days after food and fluids were terminated (December 14, 1990), Nancy Cruzan died from dehydration.

Was this a "good death" for Nancy Cruzan, an incapacitated woman who was able to chew food, take liquids, and could see and hear? Probably not. Here's one physician's description of death by dehydration: "A conscious person would feel it [dehydration] just as you or I would. They will go into seizures. Their skin cracks, their tongue cracks, their lips crack. They may have nosebleeds because of the drying of the mucus membranes, and heaving and vomiting might ensue because of the drying out of the stomach lining. They feel the pangs of hunger and thirst. Imagine going one day without a glass of water! Death by dehydration takes ten to fourteen days. It is an extremely agonizing death."[3]

The Cruzan case opened the door for killing patients who had cognitive disabilities regardless of whether they were conscious or unconscious. In fact, the AMA Council on Ethics and Judicial Affairs revised its opinion to comply with this position. It now reads: "Even if the patient is not terminally ill or permanently unconscious, it is not unethical to discontinue all means of life-sustaining medical treatment [including food and fluids] in accordance with a proper substituted judgment or best interests analysis."[4]

Many applauded death by dehydration. Dr. Kenneth F. Schaffner wrote in *Critical Care Medicine*: "It should be honestly recognized by all participants . . . [that] the right to forego food and water is the first step toward recognizing a right to medically assisted, *rational* suicide."[5] In *Rethinking Life and Death*, Peter Singer jubilantly describes his position: "The lives of such patients are of no benefit to them, and so doctors may lawfully stop feeding them to end their lives. With this decision the law has ended its unthinking commitment to the preservation of human life that is a mere biological existence. . . . In doing so they have shifted the boundary between what is and what is not murder. . . . Now, conduct intended to end life is lawful."[6]

## Compassion in Dying, Inc. v. The State of Washington

In 1991 voters in Washington State rejected a measure to legalize physician-assisted suicide. Unwilling to accept the ballot-box decision of the populace, Compassion in Dying, Inc. decided to circumvent the will of the voters by filing a 1994 suit challenging the constitutionality of the state law that prohibited assisted suicide. After a victory in Federal District Court, a three judge panel of the United States Court of Appeals for the Ninth Circuit reversed the decision. Judge John T. Noonan, writing for the court, reminded the plaintiffs that a "constitutional right to aid in killing oneself was unknown in the past," and he dismissed the argument that the right to assisted suicide was the same as the right to refuse treatment.[7]

Compassion in Dying, Inc. managed to get another hearing, this time in front of the entire eleven-member Ninth Circuit Court, and in March of 1996 the court reversed the three-member panel's earlier decision. Writing for the majority, Judge Stephen Reinhardt delivered a one-hundred-page decision which said that the law against assisted suicide violated the Fourteenth Amendment's due-process clause and that each person must be constitutionally protected in choosing "the timing and manner of one's own death."[8]

To reach his conclusion Judge Reinhardt cited the Supreme Court's decision in *Planned Parenthood v. Casey* (1992), which reads in part: "At the heart of liberty is the right to define one's own concept of existence, of meaning, of the universe, and of the mystery of human life."[9] In this view, which has come to be known among advocates of assisted suicide as the "Liberty Clause," the individual, without reference to community or tradition, has the sole power to control the meaning of life. Therefore, according to Judge Reinhardt, "when patients are no longer able to pursue liberty or happiness and do not wish to pursue life, the state's interest in forcing them to remain alive is clearly less compelling. . . . There is a constitutionally protected liberty in determining the time and manner for one's own death." [10]

The court also cited the Cruzan case, which permitted the removal of artificial life-support systems, and concluded that there is a *similarity* between what doctors are "now permitted to do and what plaintiffs assert they should be permitted to do." [11]

And there is much more in the many pages of Reinhardt's social-science sorcery which if upheld in higher courts will truly open Pandora's Box. Patients might even lose their right to choose life: "We should make it clear that a decision of a duly appointed surrogate decision maker is for all legal purposes the decision of the patient himself." [12] Death could be "hastened," even for those not terminally ill:

> There are . . . subtle concerns . . . advanced by some representatives of the physically impaired, including the fear that certain physical disabilities will erroneously be deemed to make life "valueless." While we recognize the legitimacy of these concerns, however, we also recognize that seriously impaired individuals will, along with non-impaired individuals, be the beneficiaries of the liberty interest asserted here— and that if they are not afforded the option to control their own fate, they like many others will be compelled against their will to endure protracted suffering. [13]

The decision also permits doctors to administer lethal drugs: "We recognize that in some instances, the patient may be unable to self-administer the drugs and that administration by a physician, or a person acting under his direction or control, may be the only way the patient may receive them." [14] Indeed, the decision permits most anyone to aid in suicide:

We would add that those whose services are essential to help the terminally ill patient obtain and take that medication and who act under the supervision or direction of a physician are necessarily covered by our ruling. That includes the pharmacist . . . the health care worker . . . the family member or loved one who opens the bottle, places the pills in the patient's hand, advises him how many pills to take, and provides the necessary tea, water or other liquids; or the persons who help the patient to his death bed and provide the love and comfort so essential to a peaceful death.[15]

Finances may also be a factor in choosing suicide: "While state regulations can help ensure that patients do not make uninformed, or ill considered decisions, we are reluctant to say that, in a society in which the costs of protracted health care can be so exorbitant, it is improper for competent, terminally ill adults to take the economic welfare of their families and loved ones into consideration."[16]

To rationalize the court's ruling Judge Reinhardt disingenuously refers to the Bible: "In the New Testament, the suicide of Judas Iscariot is not treated as a further sin, rather as an act of repentance."[17] This is interesting, given that Reinhardt also dismisses any religious perspective on the assisted suicide issue. "Those who believe strongly that death must come without physician assistance," he writes, ". . . are not free . . . to force their views, their religious convictions, or their philosophies on all other members of a democratic society."[18]

## VACCO V. QUILL

In a 1991 letter published in *The New England Journal of Medicine*, New York physician Dr. Timothy Quill described his participation in an assisted suicide the year before. He admitted that when he prescribed a lethal dose of barbiturates he knew he was breaking the law.

New York law permits competent adults to have living wills and health-care proxies, which may direct physicians to terminate medical treatment and life sustaining measures, but the law also prohibits assisted suicide and euthanasia. Violators can be charged with second-degree manslaughter. Section 125.15 of the New York State Penal Code states that "a person is guilty of manslaughter in the second degree when . . . he intentionally aids another person to commit suicide."

After being investigated by law enforcement officials, Dr. Quill and two medical colleagues (Samuel C. Klagsbrun, M.D. and Howard A. Grossman, M.D.) filed a suit in July 1994 challenging the constitutionality of the New York State statutes. They claimed the laws violated the due-process and equal-protection clauses in the United States Constitution. They argued that it is a constitutional right of the physician to prescribe for a mentally competent, terminally ill individual a death-producing drug that can be self-administered. They admitted that in the "regular course of their medical practice" it was within *their standards* to prescribe medicines that would hasten death in a "humane manner." In their judgment, "proper and humane medical practice should include the ability to prescribe medication which will enable a patient to commit suicide."[19]

Once again the plaintiffs employed the *Casey* Liberty Clause:

> Our law affords constitutional protection to personal decisions relating to marriage, procreation, contraception, family-relationships, child rearing, and education. . . . These matters, involving the most intimate and personal choices a person may make in a lifetime, choices central to personal dignity and autonomy, are central to the liberty protected by the Fourteenth Amendment. *At the heart of liberty is the*

*right to define one's own concept of existence, of meaning, of
the universe, and of the mystery of human life.* Beliefs about
these matters could not define the attributes of personhood
were they formed under compulsion of the State.[20] [Italics added.]

The Southern District federal court denied the complaint and
argued that the physicians' conclusions were too broad. "Suicide," it declared, "has a sufficiently different legal significance
from requesting withdrawal of treatment," therefore a fundamental right to suicide could not be inferred from Cruzan.[21] The
court also pointed out that there is no historic recognition of
suicide as a legal right; if anything, precedents point to it as a
crime. English common law, which considered suicide a criminal offense, was adopted by the American colonies, and throughout
history a majority of states have imposed criminal penalties on
someone who aids another in suicide. Hence, suicide is not a
fundamental liberty protected by the due-process clause of the
Fourteenth Amendment. The Southern District also dismissed
the argument that a physician's "refusal of treatment is essentially the same thing as committing suicide with the advice of a
physician." The court concluded it is not "irrational for the state
to recognize a difference between allowing nature to take its course
. . . and intentionally using an artificial death producing device."[22]

On April 2, 1996 a three judge panel of the Federal Court
of Appeals for the Second Circuit, in a unanimous opinion,
reversed part of the earlier decision. Judge Roger J. Miner, writing
for the court, held that physicians may prescribe drugs to mentally competent, terminally ill patients who seek to end their lives.

Although the Appeals Court declined "the plaintiff's invitation to identify a new fundamental right," it did state that the
New York statutes violated the equal-protection clause. The court
equated withdrawing life-support systems or treatment with taking prescribed drugs to hasten death. Withdrawal of life sup-

port, the court reasoned, "is nothing more nor less than assisted suicide. It simply cannot be said that those mentally competent, terminally ill persons who seek to hasten death but whose treatment does not include life support are treated equally." [23] Once again echoing the *Casey* decision, the court asked:

> What interest can the state possibly have in requiring the prolongation of a life that is all but ended? Surely the state's interest lessens as the potential for life diminishes. . . . And what business is it of the state to require the continuation of agony when the result is imminent and inevitable? What concern prompts the state to interfere with a mentally competent patient's "right to define [his] own existence, of meaning, of the universe, and of the mystery of human life," when the patient seeks to have drugs prescribed to end life during the final stages of a terminal illness? The greatly reduced interest of the state in preserving life compels this answer to these questions: "None." [24]

It appears that the court is saying that the state has a right to protect only those persons who have excellent health. In other words a person is not guaranteed protection just because he is a human being; protection now depends on the quality of his being.

According to the court, physicians

> do not fulfill the role of "killer" by prescribing drugs to hasten death any more than they do by disconnecting life support systems. Likewise "psychological pressure" can be applied just as much upon the elderly and infirm to consent to withdrawal of life-sustaining equipment as to take drugs to hasten death. There is no clear indication that there has been any problem in regard to the former, and there should be none as to the latter. . . . It is difficult to see how the relief the plaintiffs seek would lead to the abuses found in the Netherlands. [25]

But isn't this pretty much how it all got started in the Netherlands? Indeed, isn't it pretty much the way it started in Nazi Germany? One is reminded of Bishop Fulton J. Sheen's remark that what "Hitler hands out in concentrated form, we in America sell piecemeal, what Hitler sells wholesale, we sell retail."

Reactions to the Appeals Court ruling were varied:

* Plaintiff Grossman said the decision liberated "the vast underground of doctors "who have helped terminally ill people end their lives. . . . Doctors have been doing this, but they have been isolated, alone and terrified, afraid to reveal their secret event to the person they sleep next to every night."[26]

* New York State Attorney General Dennis Vacco vowed to appeal and said that in allowing physicians to help patients commit suicide the court was giving "those who swear an oath to preserve life a license to kill."[27]

* "This is an important ruling," said Dr. Samuel C. Klagsbrun [Plaintiff Physician] "in what I see as an effort to stop the encroachment, the infringement on dying patients lives—to allow them to die more peacefully."[28]

* Tracy Miller, Chairman, N.Y.S. Task Force on Life and Law, said: "It is extraordinarily dangerous to legalize assisted suicide as we rush headlong to a system of managed care. . . . It is far less expensive to assist patients in killing themselves than it is to care for them at life's end!"[29]

* "Stepping over the line [to assist suicide]," said AMA president Dr. Nancy Dickey, "is not in the best interests of patient or society."[30]

* And in his April 7, 1996, Easter Sunday sermon John Cardinal O'Connor, Archbishop of New York, said: "A Federal Court has authorized physician assisted suicide. It has taken divine law into its own hands. . . . You can't

wash your hands of it just because the law says to. . . . It will not do to say 'I am personally opposed but . . .' What makes us think that permitted lawful suicide will not become obligated suicide? How frequently will people be told to get out of the way? . . . An unspeakable decision. . . . The latest horror . . ."[31]

In October of 1996 the United States Supreme Court announced it would consider jointly the Second Circuit and Ninth Circuit cases. Twenty states urged the Court to reject the concept that terminally ill people have a constitutional right to doctor-assisted suicide. In a friend-of-the-court brief the states argued: "At stake . . . are first and foremost, the lives of the people, both those who wish to die and those who wish to live no matter what their circumstances."[32] The oral argument showdown before the Supreme Court was scheduled for January 8, 1997.

## VACCO V. QUILL IN THE SUPREME COURT

The attorney for Timothy E. Quill et al., Laurence Tribe, posed this question to the Court: "May a State constitutionally criminalize a physician's prescription of medication requested by a competent, terminally ill patient in the final states of dying, who seeks that medication to end her life without intolerable suffering, when the State permits terminally ill patients intentionally to end their lives by having their physicians withdraw or withhold essential medical treatment, including nutrition and hydration?"[33] His answer was no. Tribe then rehashed the arguments found in the lower court decision:

- As the court has made clear, the Liberty Clause protects the right to make certain "important decisions." Protection of a choice either inheres or it does not—regardless of the direction in which the individual ultimately will

seek to exercise it. With an aging population and increasingly fierce competition over medical resources, if there
is no protected liberty on the part of a dying, terminally
ill patient to decide whether or not to endure further pain
or suffering, then the State may just as readily prohibit
the provision of life-extending treatment to such patients
as it may prohibit their receipt of life-ending medication.[34]

• That liberty [for assisted suicide] follows from recognition of the basic rights to human dignity and bodily autonomy.[35]

• Indeed, the only rationale that can explain the line drawn
by New York is that the State prefers to send the (false)
message that it does not tolerate any assistance in dying,
despite the indisputable fact that physician assistance in
dying (both lawful and unlawful) in fact goes on today.[36]

• [That]. . . the liberty at issue in this case deserves constitutional protection.[37]

• The final, life-shaping choices which a dying person is
capable of making—so that, to the degree medically possible, the person's dying will be free of unbearable pain
and suffering and will not destroy the dignity and integrity of that person as a human being—are solidly encom-
passed within that liberty. Indeed, if *Casey* and this Court's
other holdings construing "the substantive sphere of liberty
which the fourteenth Amendment protects" . . . are to
be understood as principled decisions, the court must
recognize the protected nature of this profoundly personal
dimension of liberty.[38]

• Indeed, respect for the precious quality of every human
being requires that the Constitution be construed to protect
the essential dignity of those for whom life is all but over
equally with that of those with many years ahead.[39]

• Finally, the decision of one who is dying whether to endure an excruciating death involves, no less than the abor-

tion decision, "the right to define one's own concept of existence, of meaning, of the universe, and"—most profoundly—"of the mystery of human life." . . . A decision about whether to die in such a way, or instead to take some action to hasten the end of suffering—be it withdrawal of food, or self-administration of some life-ending drug—is inevitably a reflection of a person's understanding of the meaning of life and of the role of death in that life.[40]

Mr. Tribe also suggested that the statute should be eliminated because violators are not always prosecuted. In Tribe's oral arguments before the court he goes so far as to say, "The winks and nods I think affect the capacity of the system to respond humanely and rationally . . ."[41]

Attorney for the petitioners, Dennis C. Vacco, argued that there is nothing in two-hundred years of constitutional law or in the Constitution itself that suggests there is a right to assisted suicide:

◆ The right to assistance in suicide that plaintiffs claim can be found nowhere in the text of the Constitution. Its exercise has always been, and continues to be, prohibited by the great majority of the states. Nor can it be derived from either the language or the holding of *Cruzan* or *Casey*. It is, in plaintiffs' own formulation, activated by precisely the supposedly diminished quality of life of the terminally ill that is not constitutionally cognizable. And it is compelled neither by the profundity and mystery of death nor by the particular circumstances of dying in contemporary America. However the right to assistance in suicide is viewed, it is not "fundamental" and thus is not entitled to heightened protection under the Due Process Clause.[42]

• Thus, no matter what the scope of the asserted "right," whether extended only to the terminally ill or otherwise, New York may ban its exercise entirely. *Cruzan* observes that even as a state must recognize the right to refuse medical treatment, it may assert its interest in the protection of human life by prohibiting assistance in suicide. Moreover, there are many reasons to believe that the risk of error and abuse is significantly greater with assisted suicide than with withdrawal from treatment, especially under the regime of managed care that increasingly dominates the provision of health care services in America. These state interests are concededly not only rational but compelling. The difference in nature and risk between assisted suicide and withdrawal from life-prolonging treatment means that prohibiting the former while allowing the latter "is not excessive in relation to th[ose] valid purpose[s]." *Reno v. Flores* . . . Accordingly, New York's ban on assisted suicide does not violate the substantive component of the Due Process Clause.[43]

• The fact remains, however, that never once in American history has a patient lawfully obtained a physician's assistance in committing suicide. A declaration that there is a right to do so would be a leap—from near universal prohibition to constitutional obligation—unprecedented in this Court's history. There is no basis for such a leap.[44]

In a 9-0 decision, the Supreme Court reversed the Second and Ninth Circuit Courts and upheld state laws prohibiting assisted suicide. The Court rejected the expansive view of due process and equal protection.

In the State of Washington case, Chief Justice Rehnquist, writing for the court, had pointed out that the states' assisted suicide bans are not recent innovations. "Opposition to suicide," he wrote, is an enduring theme of our "philosophical, legal and cultural heritages," The Court was, therefore, "reluctant" to expand the concept of due process in "this uncharted area." In the Court's judgment, due process protects those rights and liberties that are "deeply rooted in the nation's history and tradition." The Court looks to our history and legal practices as "guideposts for responsible decision making that direct and restrain our exposition of the due process clause. . ."[45] Based on this reasoning the court concluded:

> The history of the law's treatment of assisted suicide in this country has been and continues to be one of the rejection of nearly all efforts to permit it. That being the case, our decisions lead us to conclude that the asserted "right" to assistance in committing suicide is not a fundamental liberty interest protected by the due process clause. The Constitution also requires, however, that Washington's assisted-suicide ban be rationally related to legitimate government interests. This requirement is unquestionably met here. As the court below recognized, Washington's assisted-suicide ban implicates a number of state interests.[46]

In *Vacco v. Quill* the Supreme Court ruled that New York's statutes prohibiting assisted suicide do not infringe on any fundamental rights. New York's reasoning that there was a distinction between refusing medical treatment and assisting a suicide was upheld. Also the court believed that the New York statutes had a rational public interest in mind, namely to avoid intentional killing, to maintain the physician's role as a healer, and to protect people from financial and psychological pressure.

Chief Justice Rehnquist, who delivered the opinion for the court, pointed out that the New York statues comply with the Fourteenth Amendment's equal-protection clause:

> On their faces, neither New York's ban on assisting suicide nor its statutes permitting patients to refuse medical treatment treat anyone differently than anyone else or draw

any distinctions between persons. *Everyone*, regardless of physical condition, is entitled, if competent, to refuse unwanted lifesaving medical treatment; *no one* is permitted to assist a suicide. Generally speaking, laws that apply evenhandedly to all "unquestionably comply" with the Equal Protection Clause.[47]

The argument that assisted suicide and termination of treatment are one and the same was rejected: "Unlike the Court of Appeals, we think the distinction between assisting suicide and withdrawing life-sustaining treatment, a distinction widely recognized and endorsed in the medical profession and in our legal traditions, is both important and logical; it is certainly rational."[48]

The distinction between the aggressive administering of painkilling drugs and assisted suicide was confirmed:

Furthermore, a physician who withdraws, or honors a patient's refusal to begin, life-sustaining medical treatment purposefully intends, or may so intend, only to respect his patient's wishes and "to cease doing useless and futile or degrading things to the patient when [the patient] no longer stands to benefit from them. . . ." The same is true when a doctor provides aggressive palliative care; in some cases, painkilling drugs may hasten a patient's death, but the physician's purpose and intent is, or may be, only to ease his patient's pain. A doctor who assists a suicide, however, "must, necessarily and indubitably, intend primarily that the patient be made dead."[49]

The Court concluded:

New York's reasons for recognizing and acting on this distinction—including prohibiting intentional killing and preserving life; preventing suicide; maintaining physicians' role as their patients' healers; protecting vulnerable people from indifference, prejudice, and psychological and financial pressure to end their lives; and avoiding a possible slide towards euthanasia—are discussed in greater detail in our opinion in *Glucksberg, ante*. These valid and important public interests easily satisfy the constitutional requirement that a legislative classification bear a rational relation to some legitimate end.[50]

As one would expect, the pronouncement was cause for celebration in some circles while in others there were howls of despair. The Orthodox Jewish organization, Aqudath of America, called the ruling "a victory for the principle that life should be measured by its inherent sanctity, not by its perceived quality."[51] The National Catholic Conference agreed, and stated: "True compassion for the dying is not shown in making them die, but in providing aid, palliative care and the comfort of human concern."[52] Steven Shapiro, head of the American Civil Liberties Union, stated that Americans "should have the right to die in a humane and dignified manner."[53] And asked how the decision would affect his "medical" practice, Dr. Jack Kevorkian said it would make "not one damn bit of difference."[54]

The Court's decision in *Vacco* does not end this discussion of assisted suicide. Proponents were quick to point out that the Court did not seal the door shut.

In a concurring opinion in *Washington v. Glucksberg* (1997), Justice David Souter wrote: "While I do not decide for all time that respondents claim should not be recognized, I acknowledge the legislative institutional competence as the better one to deal with that claim at this time."[55] Justice John Paul Stevens also left a crack in the door with his observation that the court's holding "does not foreclose the possibility that some applications of the New York statute may impose an intolerable intrusion on the patient's freedom."[56] Agreeing with his colleagues, Chief Justice Rehnquist put it this way: "Throughout the nation, Americans are engaged in an earnest and profound debate about the morality, legality, and practicality of physician-assisted suicide. Our holding permits this debate to continue, as it should in a democratic society." [57]

# 14

# DOES A PERSON HAVE THE "RIGHT TO DIE"? THE DEBATE GOES ON

> *Not only is suicide a sin. It is the ultimate and ab-*
> *solute evil, the refusal to take an interest in exist-*
> *ence; the refusal to take the oath of loyalty to life.*
> *The man who kills a man, kills a man. The man*
> *who kills himself, kills all men; as far as he is*
> *concerned he wipes out the world. . . . When a man*
> *hangs himself on a tree, the leaves might fall off*
> *in anger and the birds fly away in fury; for each*
> *has received a personal affront. . . . There is a*
> *meaning in burying the suicide apart. The man's*
> *crime is different from other crimes—for it makes*
> *even crimes impossible.*
>
> G.K. Chesterton

WHEN A PERSON takes a position on the "right to die" issue, he is also embracing, whether he knows it or not, one or another of the theological or philosophical arguments that support the position. A citizen, legislator, or judge must define liberty, justice, human nature, and the right to self-determination in order to reach any conclusion on this issue, and any suggestion that one can be morally neutral in choosing sides about assisted suicide and euthanasia is absurd.

Those who support a right to die usually assume that the right to life is neither natural nor inalienable, and that one has an absolute right to freely and privately assess the value of one's life and—if the quality of life is not satisfactory—to choose death. In their judgment life can be unwanted because the person has privately determined that he can no longer achieve, exercise, or enjoy his personal values or is unhappy with his current state.

This position is based on the utilitarianism of Jeremy Bentham and John Stuart Mill. As we noted in chapter five, Mill concluded that "the creed which accepts as the foundations of morals, utility or the greatest happiness principle holds that actions are right in proportion as they tend to promote happiness, are wrong as they tend to promote the reverse of happiness. By happiness is intended pleasure and the absence of pain; by unhappiness, pain and the privation of pleasure."[1] In other words only the useful or the enjoyable—that which has utility value for the person— is good. This was in accord with Mill's teacher Bentham, who wrote that "pleasure is in itself good, nay, even setting aside immunity from pain, the only good: pain is in itself evil; and indeed, without exception the only evil; or else [the] words good and evil have no meaning."[2]

The moral doctrine that the good is based on sensual pleasure of the individual is not new. It pops up throughout the history of philosophy. Democritus (460-370 BC) taught that attaining ones desires is the highest good. Aristippus (435-354 BC) believed that the pleasure of the body is the supreme good. Epicurus (341-270 BC) held that pleasure is the only good and pain the only evil. For him, life devoid of pleasure is without value; he insisted total control over his life when suffering. The Epicurean poet Lucretius wrote: "If one day, as well may happen, life grows wearisome, there only remains to pour a libation to death and oblivion. A drop of subtle poison will gently close your eyes to the sun, and waft you smiling into the eternal night whence everything comes and to which everything returns."[3]

Pico della Mirandola (1463-1494) in his work, *Oration of the Dignity of Man*, argued that suicide is a form of human dignity. Such seventeenth- and eighteenth-century writers as John Donne, David Hume, Voltaire, Jean-Jacques Rousseau, and Baron de Montesquieu all believed that when life does not give pleasure, when it becomes intolerable and devoid of benefits, suicide is both justified and rational.

Modern champions of suicide such as Jack Kevorkian, Derek Humphry, and Joseph Fletcher subscribe to the utilitarian position. They too argue that one must have total control over one's life and that in times of pain choosing death is virtuous and should be brought on "sweetly," "gently," "softly," "silently," and "mechanically."

In the 1986 case, *Brophy v. New England Sinai Hospital*, Judge Liacos, writing for the Massachusetts Supreme Court, ruled that Paul Brophy (who was in a semi-vegetable state) had the right to self-determination and that the state could not interfere or deny his desire to die. To justify the court's position Justice Liacos quoted Mill: "[The] only purpose for which power can be rightfully exercised over any member of a civilised community, against his will, is to prevent harm to others. His own good, either physical or moral, is not a significant warrant. He cannot rightfully be compelled to do or forbear because it will be better for him to do so, because it will make him happier, because, in the opinion of others, to do so would be wise or even right."[4]

To avoid pain the patient has the right to choose death over life. According to this court decision, Mr. Brophy had the right to privacy and autonomy over his body. Dying has the same constitutional protection as living, and, so long as no one else is harmed, justice is served when the person makes a decision concerning his body. Pleasure in this case means choosing to die with dignity. Justice is not based on the good of the community or any absolute standards of right and wrong, but on the individual's choices concerning the quantity of life. The sole criteria is based on the individual's private assessment—on the eye of the beholder.

In *Bouvia v. Superior Court* (1986), a California case that involved the removal of life-sustaining treatment, the California Court of Appeals adopted the utilitarian view with this comment: "It is incongruous, if not monstrous, for medical practitioners to assert their right to preserve a life that someone else must live, or more accurately, endure for '15 to 20 years.' We cannot conceive it to be the policy of this State to inflict such an ordeal on anyone." [5]

The court decreed that the mentally competent patient in question, who was not terminally ill, "has the right to have [the medical device] removed immediately. . . . The exercise of [that right] required no one's approval. . . . It was a moral and philosophical decision that, being a competent adult was hers alone. Her human dignity rests on her right to make decisions based on her values or plans. She has sovereignty over her life and can self-determine her fate." [6]

Although the California court in this decision did not overturn the ban on assisted suicide, it certainly came close to doing so: "The right to die is an integral part of our right to control our own destinies so long as the rights of others are not affected. That right should, in [my] opinion, include the ability to enlist assistance from others, including the medical profession, in making death as painless and quick as possible."[7]

As one of the justices wrote in a concurring opinion, "If there is ever a time when we ought to be able to get the ' government off our backs,' it is when we face death—either by choice or otherwise." The medical profession was criticized for arguing that there was value in living, that life was not devoid of meaning. The value of life, according to the Court, could be rendered meaningless and the person possessed the fundamental right to choose death. This view was also voiced by Supreme Court Justice Stevens in the Nancy Cruzan case: "the sanctity of the individual's privacy of the human body is obviously fundamental to liberty."[8]

But is the individual's total privacy and right to die a fundamental liberty? Is the Mill-Bentham utilitarian moral and philosophical approach the basis of our constitutional government? Is the person autonomous?

Part I of this book demonstrated the necessity of government based on the common good. As social beings, people by their nature come together to form families, communities, states, and nations. The common good of the community challenges the absolute sovereignty of the individual. Public welfare is structured to protect and support the dignity of all its members. Hence government has an obligation to assess the effects of every public policy decision on the society as a whole. Government must listen to and respect the concerns of all individuals, but ultimately a decision based on the common good must supersede the private interests of any individual. The decision must contribute to the well being of the entire community.

There are times when the wishes, desires, and so-called rights and actions of individuals will be limited to promote the general welfare. Most of our laws, criminal codes, traffic laws, and housing and fire codes are based on this premise. Maintaining the sanctity of life—the inalienable right to life—is a fundamental basis of our rule of law.

Thomas Hobbes and John Locke, whose views influenced the Founding Fathers, argued that the natural drive of a person is to preserve life. People form governments to protect that basic urge and no government can take away that right—it can never be surrendered to the sovereign. Judeo-Christian traditions also affirm the state's inherent duty to preserve the sanctity of all human life.

In the *Brophy* case, Judge Lynch, in his dissenting opinion, acknowledged that "the state's interest in the preservation of life has not been given appropriate weight."[9] Judge Nolan, another dissenter, recognized the moral and philosophical arguments at play and stated the majority view was "another triumph for the forces of secular humanism (modern paganism)."[10] Judge O'Conner in a partial dissent, pointed out that the court's rationale for establishing suicide as a legal right is contrary to "this nation's traditional and fitting reverence for human life." He continued: "Even in cases involving severe and enduring illness, disability and 'helplessness,' society's focus must be on life, not death, with dignity. By its very nature, every human life without reference to its condition, has a value that no one rightfully can deny or measure. Recognition of that truth is the cornerstone on which American law is built."[11]

In the *Brophy* case, the court split over conflicting moral and philosophical approaches; over utilitarianism versus the common good.

Is state-sanctioned assisted suicide contrary to the common good? Can it endanger the welfare of the entire community?

Shortly after the Second and Ninth Circuit Courts announced their decisions declaring the ban on assisted suicide unconstitutional, stories in the *New York Times* contained these headlines:

"The Right to Suicide, Some Worry, Could Evolve into a Duty to Die" (April 7, 1996); "Suicide Ruling Raises Concern: Who Decides?" (April 4, 1996); "Concerns Grow that Doctor-Assisted Suicide Would Leave the Powerless Vulnerable—Some say expanding the power of life and death could lead to a slippery slope" (October, 20, 1996). As these headlines imply, the right to assisted suicide involves a series of issues that could violate the general welfare of the citizenry of this nation.

In 1994, the New York State Task Force on Life and the Law published a volume titled: *When Death is Sought: Assisted Suicide and Euthanasia in the Medical Context.* This finding (praised in medical and legal circles and cited as an authoritative source in the Supreme Court's decision in *Vacco v. Quill*) addressed the concerns raised in the *New York Times* articles. While the members of the task force represented all sides of the assisted suicide debate, their conclusion that New York State's laws banning the procedure should not be eliminated was unanimous. The report states that "the Task Force members unanimously concluded that legalizing assisted suicide and euthanasia would pose profound risks to many patients . . . positing an `ideal' or `good' case is *not* sufficient for *public policy* if it bears little relation to prevalent social and medical practices."[12] [Italics added.] In other words the report reasoned both that the common good overrides the private interest and that liberty and justice are best served by not eliminating the ban on assisted suicide. Their case is compelling and worth summarizing:

- Regardless of guidelines, the practices "will pose the greatest risk to those who are poor, elderly, members of a minority group or without access to good medical care."

- The threat of ever increasing health care costs "increase[s] the risks [and pressures] presented by legalizing assisted suicide and euthanasia."

- Proposed safeguards would not "prevent abuse and errors."

- While there is deep compassion for those "rare" cases when it is impossible to curtail pain, as a society "we have better ways to give people greater control and relief from suffering than by legalized assisted suicide and euthanasia."

- Those who choose suicide are generally in a state of depression. "Even if diagnosed, depression is often not treated. In elderly patients as well as the terminally and chronically ill, depression is grossly underdiagnosed and undertreated."

- Unrelieved pain often causes a person to turn to suicide. "The undertreatment of pain is a widespread failure of current medical practice, with far reaching implications for proposals to legalize assisted suicide and euthanasia."

- Legalization will "blunt our perception of what it means . . . to take another person's life."

- "The criteria and safeguards . . . would prove elastic in clinical practice and law. . . . Euthanasia [for] . . . those who are incapable of consenting would also be a likely, if not inevitable, extension of any policy permitting the practice for those who cannot consent."

- "These concerns are heightened by experience in the Netherlands . . . [where] 1,000 deaths occurred [annually] without a specific request. Moreover, in some cases, doctors have provided assisted suicide in response to suffering caused solely by psychiatric illness including severe depression."[13]

The elimination of the assisted suicide ban would also be an assault on the underlying values of the medical profession. In his 1941 sermon condemning Nazi euthanasia, Bishop von

Galen raised this question: "who could trust his physician?" He was right. The German and Dutch experiences have resulted in the destruction of the doctor-patient relationship. Legalized assisted suicide in the United States could have the same effect and the integrity of our medical services could be jeopardized.[14] The physician's calling is to cure and comfort the patient. Doctors are expected to relieve those forms of suffering that medically accompany serious illness and the threat of death. They should relieve pain, allay anxiety, and be a comforting presence.[15] To compel a physical to participate in an assisted-suicide decision goes beyond his field of expertise. The doctor would be called upon to make moral and philosophical decisions on the value of a person's life. Noted scientist-philosopher Leon Kass believes that putting doctors in the role of physician-euthanizers is oxymoronic. He wonders if one can "benefit the patient as a whole by making him dead? There is of course, a logical difficulty: how can any good exist for a being that is not? But the error is more logical: to intend and to act for someone's good requires his continued existence to receive the benefit."[16]

If the physician is permitted to serve as a public executioner, he can be susceptible to pressures that do not benefit patients. Hospital administrators, social workers, and a patient's relatives and friends could influence a doctor's decision. Finances and family pressures could become the basis of life-and-death decisions, and patients could become the victims of animosity and greed. If the character of medicine includes killing, the patient may no longer trust the motives of the doctor. The patient may lose all faith in a profession that is not solely dedicated to restoring or maintaining health. In Germany the medical profession's character was perverted; doctors saw themselves as more responsible for the health of the nation than for the good of the individual patient. This view now permeates American medicine. In a March 18, 1994 segment of the *McNeill-Lehrer Newshour*, Dr. Donald Murphy, a geriatric physician at Denver's St. Luke Hospital, stated: "The whole question about futile care [which can be an erroneous judgment] or inappropriate care was an issue 10 years ago. But it didn't have the urgency it does today. The reason it has the urgency today is because of health care reform. A lot of us feel that if we're really going to have a just health care system, we're going to have to set some boundaries on care at the margins. If we want to bring everyone into the fold and give fair care to everyone, we're going to have to set some limits where care is marginal."[17]

Dr. Herbert Hendin, Executive Director of the American Suicide Foundation, after an extensive analysis of the Dutch euthanasia experience confirmed that the practice of medicine has changed dramatically: "Acceptance of euthanasia in the Netherlands has reduced interest in alleviating pain and suffering; euthanasia becomes an easier alternative—even when a person is not terminally ill."[18]

If assisted suicide gains acceptance as a medical treatment it can be expected to destroy the trust patients have in their doctors; to undermine the expectation that the physician's commitment is always to life. It would almost surely introduce a utilitarian approach to medicine that violates the American conscience and the state's interest to preserve the common good.

In his extraordinary work *The Ethics of Rhetoric,* philosopher Richard Weaver wrote; "All things considered, rhetoric, noble or base, is a great power in the world; and we note accordingly that at the center of the public life of every people there is a fierce struggle over who shall control the means of rhetorical propagation."[19] The German, Dutch, and American proponents of assisted suicide formulate their arguments in terms intended to soothe and exploit their intended victims. These words and phrases appear again and again when describing suicide: beautiful, merciful, compassionate, kind, caring, charitable, ethical, honorable, loving, dignified, humanitarian, precious, healing treatment, healing work, rational suicide, balanced suicide.

These adjectives and expressions are a charade employed by those who wish to advance the culture of death. The deliberate taking of innocent life which they advocate is a violation of the natural law and is contrary to the common good. If the state grants the power to take even one innocent life it opens Pandora's Box. Once the principle of the sanctity of life, the inalienable right to life, is breached, one can rationalize the taking of life for any reason—economic, racial, religious, or any other.

This does not mean, however, that as a nation we can ignore the plight of the terminally ill. In his work *Seduced by Death*, Dr. Herbert Hendin puts it this way:

> Euthanasia advocates have been seduced by death. They have come to see suicide as a cure for disease and a way of appropriating death's power over the human capacity for control They have detoured what could be a constructive effort to manage the final phase of life in more varied and individualistic ways onto a dangerous route to nowhere. These are not the attitudes on which to base a nation's compassionate social policy. That policy must be based on a larger and more positive concern for people who are terminally ill. It must reflect an expansive determination to relieve their physical pain, to discover the nature of their fears, and to diminish suffering by providing meaningful reassurance of the life that has been lived and is still going on.[20]

In order to live with dignity until the moment of natural death, a chronically ill, disabled or dying person—like any human being—has the right to compassionate, humane, and medically indicated treatment and care. Health officials have an obligation to promote and make accessible proper palliative care.[21]

- Doctors must be trained to diagnose and treat depression in the elderly and chronically ill.

- Health professionals must be trained in the latest pain-control techniques. Patients must be informed of the benefits and have access to medicines and therapies which suppress pain.

- When it is no longer possible to cure, health officials must be trained to care for the terminally ill.

- Patients must be informed about hospice care. The *Harvard Health Letter* states: "The Hospice philosophy is that dying should be accepted as a unique part of life, not resisted with every weapon in medicine's armamentarium. When nothing more can be gained from [curative] treatment, hospices focus on making people as comfortable as possible."[22]

These are just a few of the measures that must be implemented to ensure that proper care is provided for severely ill patients. The American Foundation for Suicide Prevention, The Compassionate Health Care Network, National Chronic Pain Outreach Association, The National Hospice Organization and many other groups are dedicated to promoting a wide range of programs to improve the state of care for the ill and disabled.

The culture-of-death creed would have us abandon the sick and elderly. Americans, whose most basic belief is in the inalienable right to life, have an obligation to promote a culture of compassion, one which ensures that every person lives—every moment of his life—with dignity.

PART III

# EUGENICS AND
# CLONING

# 15

## THE SCIENCE OF GOOD BIRTH

*[E]ugenics is chiefly a denial of the Declaration of Independence. It urges that so far from all men being born equal, numbers of them ought not to be born at all. And so far from their being entitled to life, liberty, and the pursuit of happiness, they are to be forbidden a form of liberty and happiness so private that the maddest inquisitor never dreamed of meddling with it before.*

G.K. Chesterton

IN HIS 1869 BOOK, *Hereditary Genius,* Sir Francis Galton (1822-1911) proposed eliminating society's less desirable individuals and multiplying its more desirable members. He was confident that a highly gifted race of men and women could be produced by means of judicious marriages during several consecutive generations,[1] and he hoped the unworthy could be comfortably segregated in monasteries and convents where they would be denied the right to propagate.[2] This was the inception of the modern eugenics movement; the so-called "science of good birth."

Although these ideas flourished in the nineteenth century, the ideological origins of eugenics may be traced back to the advent of Cartesian reductionism, which measures the universe (mankind included) in exclusively material terms. As Descartes' earlier contemporary Johannes Kepler (1571-1630) put it: "Nothing can be known except quantitatively."[3]

Aristotle's famous "hylomorphism," which taught that man is composed of prime matter and substantial form (body and soul), was totally rejected by the materialists. Man is soulless, they insisted, not made in the image and likeness of God, and the supposed difference between man and beast, between for instance a saint and a rodent, is merely a matter of degree.

This is completely alien to the classical definition of the person which defines the person as a "complete individual of intellectual nature" (Boethius). This view includes God and the angels with man, and, being non-reductive, excludes rodents. The *sine qua non* of the person is the soul, which is not simply an extension of a material or animal nature. It was this unity of body and soul that led Thomas Aquinas to say that the human person "signifies what is most perfect in nature."[4] With a soul, man has nobility. Soulless man is without dignity.

For Descartes and the later rationalists a man is not a person, he is a thing. "There is no difference between cabbages and kings," Nobel laureate Albert Szent Gyorgyi once quipped. "We are all recent leaves on the old tree of life."[5] This is very much the vision of evolution as expressed by Francis Galton's cousin Charles Darwin (1809-1882) in *The Origin of the Species by Means of Natural Selection on the Preservation of Favored Races in the Struggle for Life.* (It is interesting to note that most of book's title is usually dropped by those who cite it. Could this be because the word "favored" exposes the book's eugenic implications?) According to Darwin nature constitutes a continuum among organisms which are not essentially diversified (one animal is pretty much like another), but are fiercely competitive. Competition among and within the species leads to "war," because inequality is a law of nature.[6] He states that he "could show [that war had] done and [is] doing [much] . . . for the progress of civilization. . . . The more civilized so-called Caucasian races have beaten the Turkish hollow in the struggle for existence. Looking to the world at a not very distant date . . . an endless number of the lower races will have been eliminated by the higher civilized races throughout the world."[7]

In speaking of the elimination of the "lower races," Darwin makes a frank avowal of eugenics: "We civilized men, do our utmost to check the process of elimination; we build asylums for the imbecile, the maimed and the sick; we institute poor laws; and our medical men exert their utmost skill to save the life of every one to the last moment . . . Thus the weak members of civilized societies propagate their kind. No one who has attended to the breeding of domestic animals will doubt this must be highly injurious to the race of man."[8]

In expounding his muscular W*eltanschauung*, German philosopher Friedrich Nietzsche (1844-1900) applied the theory of biological evolution to the wider context of cultural development[9] and concluded that the ancient belief in the sanctity of the individual had been replaced by a new creed of the survival of the fittest:

Society as the trustee of life is responsible to life for every botched life that comes into existence; and as it has to atone for such lives, it ought consequently to make it impossible for them ever to see the light of day; it should in many cases actually prevent the act of procreation, and may, without any regard for rank, descent, or intellect, hold in readiness the most rigorous forms of compulsion and restriction, and, under certain circumstances, have recourse to castration. . . . "Thou shalt do no murder," is a piece of ingenuous puerility compared with "Thou shalt not beget"!!! . . . The [unhealthy] must at all costs be *eliminated*, lest the whole fall to pieces. [Italics in original.][10]

According to Herbert Spencer (1820-1903), the quality of the human species improves slowly through the process of evolution and cannot be changed for the better by any other means. In primitive conditions, he argued, people necessarily resort to violence and warfare for survival, and war has the positive effect of killing off inferior races, and this promotes the survival and reproduction of superior human strains. In complex societies the conflicts among people become more economic than military, and Spencer warned the developed nations not to foster the survival of the unfit by interfering with harsh, but essential, economic realities.[11] In the name of biology he opposed free public education, sanitation laws, compulsory vaccinations, and welfare programs for those he called the "hereditary poor." He feared that the utilization of these services would encourage the perpetuation of undesirable physical, intellectual, and social traits. Spencer's version of what would become Social Darwinism made the pseudo-science of eugenics "morally" permissible in the name of preserving "society as a whole."[12]

The belief that the evolution of the human race may be improved by programs of breeding which foster more desirable traits than nature alone may provide is called eugenics or *positive* eugenics.[13] *Negative* eugenics (also known as dysgenics or cacogenics) would "purify" the gene pool by breeding out undesirable traits or by disposing of undesirable human beings: individuals, ethnic groups, or whole races.[14]

British scientist (and Galton disciple) Karl Pearson declared that "no training or education can create intelligence, you must breed it."[15] In his book *Ethics of Free Thought*, Pearson warned that "the social imperialist state might well have to intervene in reproductive matters at least in the families of anti-social propagators of unnecessary human beings."[16] Of course this begs the question: Who exactly are the "propagators of unnecessary human beings"? Probably the poor. Perhaps you and I. Like all gnostics, ancient and modern, who claim to know the path which will lead man to earthly perfection, Pearson and other eugenicists are certain that they can provide the leadership necessary to identify undesirables and then implement  plans to "intervene" in their sex lives.

Great Britain has produced a number of public figures who have been outspoken in favor of eugenics. Playwright George Bernard Shaw declared that there "is now no reasonable excuse for refusing to face the fact that nothing but a eugenics religion can save our civilization from the fate that has overtaken all previous civilizations."[17] Author H.G. Wells claimed that the children are "no more . . . their [parents'] private concern entirely than the diseases germs disseminate or the noises a man makes in a third floor flat."[18] Political philosopher Harold Laski (another Galton protégé) wrote in a letter to Justice Oliver Wendell Holmes in which he urged sterilization of "all the unfit, among whom I include all fundamentalists."[19] And noted socialist Sydney Webb wrote: "In Great Britain at this moment, when half, or perhaps two-thirds of all the married people are regulating their families, children are being freely born to the Irish Roman Catholics and the Polish, Russian, and German Jews . . . and the thriftless and irresponsible. . . . This can hardly result in anything but national deterioration. . . ."[20]

The situation in America was little different. According to Temple University professor Mark Haller: "[At the turn of the century] educated Americans came increasingly to identify themselves and their values with the Anglo-Saxon race . . . and its love for liberty. That same love of liberty caused the peoples of Northern Europe to accept Protestantism while more servile people of southern Europe remained under the domain of Rome."[21]

To northeastern progressive Republicans (the Rockefellers, for instance), the Irish, Italian, and Eastern-European immigrants who settled in America's cities were distasteful if not repulsive. In New York, the Reverend Frank Marling of the Second Avenue Presbyterian Church declared: "The vast hordes flocking to this land strike at our national life, which we count most precious, while the ballot gives them power which they know too well how to use."[22] Margaret Sanger, one of contemporary feminism's greatest heroines, said in 1921 that eugenics is "suggested by the most diverse minds as the most adequate and thorough avenue to the solution of racial, political and social problems. The most intransigeant [*sic*] and daring teachers and scientists have lent their support to this great biological interpretation of the human race." Sanger, founder of Planned Parenthood, boldly championed "more children for the fit, less from the unfit, that is the chief issue of birth control."[23]

Theodore Roosevelt and Henry Cabot Lodge, the two men who epitomized the progressive movement, mourned what Madison Grant called the "passing of the great race" and embraced Social Darwinism to rationalize their Anglocentrism.[24] Influenced by eugenicist Brooks Adams, Roosevelt called for the sterilization of criminals and scolded the upper classes for committing "race suicide."[25] (He was upset that Harvard graduates were producing only one-half to two-thirds of their original number.)

Shortly before leaving the White House, Roosevelt wrote to President-elect William Howard Taft:

> Among the various legacies of trouble which I leave you there is none as to which I more earnestly hope for your thought and care than this. There are very big problems which we have to face in the United States. I do not know whether you yourself realize how rapid the decline in the birth rate is, how rapid the drift has been away from the country to the cities. In spite of our enormous immigration, there is a good reason to fear that unless the present tendencies are checked your children and mine will see the day when our population is stationary, and so far as the native stock is concerned is dying out.[26]

Once he left office Roosevelt took an active role in the debate about eugenics. "Someday," wrote Roosevelt in 1913 to Charles Davenport, director of the Eugenics Record Office, "we must realize that the prime duty, the inescapable duty of the good citizen of the right type is to leave his or her blood behind him in the world, and that we have no business permitting the perpetuation of citizens of the wrong type." He continued:

> Any group of farmers who permitted their best stock not to breed, and let all the increase come from the worst stock, would be treated as fit inmates for an asylum. Yet we fail to understand that such conduct is rational compared to the conduct of a nation which permits unlimited breeding from the worst stocks, physically and morally, while it encourages or connives at the cold selfishness or the twisted sentimentality as a result of which the men and women who ought to marry, and if married have large families[,] remain celibates or have no children or only one or two.[27]

To implement these views Senator Henry Cabot Lodge introduced draconian legislation (sponsored by the Immigration

Restriction League) and denounced his congressional colleagues for not understanding the threat posed by the "immigration of people of those races far removed in thought and speech and blood from the men who have made this country what it is."[28]

Other examples of prominent turn-of-the-century Americans who advocated eugenics include:

- John Humphrey Noyes, founder of the Perfectionist Oneida Community in upstate New York, who frowned upon monogamy, which to him discriminated against the best in favor of the worst. The good man, limited by his conscience, will have few offspring, Noyes believed, while the bad man, free from moral restraint, will reproduce abundantly.[29]

- The Carnegie Institute endowed ten million dollars in 1904 for a "eugenics station" at Cold Spring Harbor on the north shore of Long Island. Director Charles Davenport believed behavior is genetic and determined by race. He looked forward to a day when a woman would no more marry a man without knowing his genetic history than a horse breeder would acquire a stallion without knowing its pedigree.[30] In a 1910 letter to Sir Francis Galton, Davenport wrote, "We cannot ask all persons with a defect not to marry, for that would imply *most* people. . ."[31]

- Philanthropist Mary Harriman, sister of the governor of New York, financed the Eugenics Record Office, the purpose of which was to gather hereditary information about the relatives of those deemed defective. Objections to its operations on the grounds of invasion of privacy were discounted as "a narrow and false view."[32]

- John D. Rockefeller, Jr. (son of the founder of Standard Oil) and George Eastman (founder of Eastman Kodak) financed the American Eugenics Society. In 1926 the Society's Eugenics Sermon Contest was won by the Rev.

Dr. Kenneth MacArthur, of Sterling, Massachusetts, who sermonized that moral and spiritual qualities—intelligence for instance—were hereditary.[33]

The interest in eugenics among America's best and brightest Victorians led throughout America to the proposition of laws which called for restrictions on childbearing and for sterilization of a number of groups in society. People genuinely believed that unrestricted "breeding" by blacks, by the poor, and by immigrants threatened the dominion of the "civilized races."

In 1922 musician John Powell founded the Anglo-Saxon clubs of America. A *Richmond* (Virginia) *News Leader* article outlined the group's goals: "The fundamental purpose of the organization is the preservation and maintenance of Anglo-Saxon ideals and civilization in America. This purpose is to be accomplished in three ways: first, by the strengthening of Anglo-Saxon instincts, traditions and principles among representatives of our original American stock; second, by intelligent selection and exclusion of immigrants; and third, by fundamental and final solutions of our racial problems in general, most especially of the Negro problem."[34]

In 1926 the American Eugenics Society issued *A Eugenics Catechism*, which it presented in the classic question-and-answer format:

Q: Does eugenics contradict the Bible?
A: The Bible has much to say for eugenics. It tells us that men do not gather grapes from thorns and figs from thistles . . .

Q: Does eugenics mean less sympathy for the unfortunate?
A: It means a much better understanding of them, and a more concerted attempt to alleviate their suffering, by seeing to it that everything possible is done to have fewer hereditary defectives . . .

Q: What is the most precious thing in the world?

A: The human germ plasm.[35]

And in 1927 the United States Supreme Court weighed in with its decision in *Buck v. Bell.* The plaintiff, Carrie Buck, had been forcibly sterilized at the Virginia State Colony for Epileptics and Feeble-minded, and her case became a test of the constitutionality of the state's sterilization law. Writing for the majority, Associate Justice Oliver Wendell Holmes made his infamous comment that "three generations of imbeciles are enough." He elaborated:

> We have seen more than once that the public welfare may call upon the best citizens for their lives. It would be strange if it could not call upon those who already sap the strength of the State for these lesser sacrifices, often felt to be much by those concerned, in order to prevent our being swamped with incompetence. It is better for all the world, if instead of waiting to execute degenerate offspring for crime, or to let them starve for their imbecility, society can prevent those who are manifestly unfit from continuing their kind. The principle that sustains compulsory vaccination is broad enough to cover cutting the Fallopian tubes . . .[36]

At the beginning of World War II thirty states had enacted laws which called for compulsory sterilization of the poor, the helpless, and the misdiagnosed.[37] The majority of these laws were based on the Model Eugenical Sterilization Law, drafted by Harry H. Laughlin, superintendent of the Eugenics Record Office.[38] Laughlin's law called for the sterilization—"regardless of etiology or prognosis"—of criminals, mental patients, the feeble-minded, inebriates, the blind, the diseased—which included those with serious physical impairments—the deaf, the deformed, and the dependent, that is, the homeless, orphans, tramps, and paupers.[39]

In his work *Preface to Eugenics* (1940) Frederick Osborne of the American Museum of Natural History called for the segregation and training of the "hereditary defective" in state institutions. Inmates were to be paroled only after sterilization. He further advocated that normal people who were known to be carriers of hereditary disabilities be sterilized, and that those with any family history of defects be encouraged to use contraceptives.[40] Although Osborne paid lip service to democratic principles, he could not hide his elitism. "It is doubtful," he wrote, "whether democracy can long continue in any society ex-cept one whose operation favors the survival of competent people in every social and occupational group."[41]

# 16

# THE GERMAN EUGENICS EXPERIENCE

> *The* Völkisch *state must see to it that only the healthy beget children. . . . Here the state must act as the guardian of a millennial future. . . . It must put the most modern medical means in the service of this knowledge. It must declare unfit for propagation all who are in any way visibly sick or who have inherited a disease and can therefore pass it on.*
>
> Adolph Hitler

THE NATIONAL SOCIALISTS were the first to make eugenics a matter of public policy. In *Mein Kampf* Hitler had declared that the task of the German people was to be "assembling and preserving the most valuable [Aryan] stocks... slowly and surely raising them to a dominant position."[1] Joseph Goebbels ordered all German organizations to be educated in "the eugenics way of thinking."[2]

Hitler became Chancellor in January of 1933 and within months eugenics legislation was being signed into law:

- April 7, 1933: A new civil-service law requires proof of Aryan ancestry as a qualification for government employment.

- July 14, 1933: A sterilization law is enacted.

- September 29, 1933: New rules mandate that inheritors of family farms must be of German blood.

- November 24, 1933: A castration law is passed.

- September 15, 1935: The Blood Protection Law bars marriages between Germans and Jews.

- October 18, 1935: The Marital Health Law requires a certificate of "genetic purity" from potential marriage partners.

- November 15, 1935: New citizenship laws deprive Jews of civil rights.[3]

This racial legislation provided the legal foundation for programs that affected millions of innocent human and that utilized euthanasia, artificial insemination, abortion, electric-shock experiments, tissue and muscle experiments, and fetal experimentation.[4] All these horrors took place in the name of eugenics.

And while the Holocaust has laid bare the Nazi malevolence towards Jews, it must not be forgotten that Hitler and his new order eagerly sought to use eugenics to rid the world of a number of other groups, not least among them the Poles, one of the Slavic "races," whom the Nazis considered sub-human. In no country conquered by German forces was repression in the name of eugenics more terrible than in Poland. The Nazis aborted the fetuses of pregnant Slavic workers, outlawed sexual intercourse between Poles, and, most dramatically, covered the entire country with a network of two thousand concentration camps. Jews, from Poland and nations throughout Europe, died mostly in the camps' gas chambers, while the majority of Christian Poles perished in mass or individual executions or were starved or worked to death.[5] Of the six million Polish citizens liquidated by the Nazis, fifty percent were Polish Jews and fifty percent were Polish Christians. Fully ninety percent of Polish deaths during the war were non-military casualties.[6]

The blame for these crimes is usually laid at the feet of the warped ideology of demented Nazi thugs. But this view ignores the fact that many of the scientists who constructed the theoretical bases for Nazi eugenic practices were highly esteemed; and not only in Germany. In America, for example, Carlos Closson of the University of Chicago professed to be able to judge racial and individual qualities by using the Retzius-Broca "cephalic index," a cockeyed set of measurements of a person's cranium.[7] Others who supported the cephalic index as a tool of scientific racism included M.I.T. sociologist William Ripley (author of *Races of Europe*, 1899), Carl C. Brigham of Princeton University, the aforementioned Karl Pearson (University College, London), and the pathologically anti-Semitic American, Madison Grant, founder of the white-supremacist Pioneer Fund, which exists to this day.[8] (Grant actually argued that infanticide was a law of nature which was unjustly and unwisely circumvented because of mere "sentimental [Christian] belief.")[9]

On the Continent, Count Georges Vacher de Lapouge published *L'Aryen: son rôle social*, in which he argued that Poles, Jews, and, for that matter, Catholics were brachycephalic (round-headed), an inferior racial type and a "proliferating menace."[10] The eugenic justification for the destruction of Jews, Poles, Slavs, Italians, and others considered "unfit" (including infants), had been on the scene well before being implemented by the Nazis.

Indeed there is ample evidence that German scientists during this period were greatly influenced by the American eugenics movement.[11] The books of eugenicist Lothrop Stoddard, a Harvard Ph.D., and Madison Grant found a "large circle of readers," and played a significant role in awakening in Germany "the movement for the preservation and increase of the Nordic race."[12] Americans eugenicists, men such as Davenport and Laughlin of the Eugenics Record Office, were very impressed—even envious—about the thoroughness with which racial measures were administered when the Nazis opened the first Race Bureau for Eugenic Segregation and the Eugenics Courts in 1933. One anonymous employee at Cold Spring Harbor remarked that Hitler "should be made an honorary member of the Eugenics Record Office."[13] When Lothrop Stoddard was a visiting professor of eugenics at a German university in 1940, he actually sat on the bench of the German Eugenics Supreme Court. (This court had the power to order sterilization of virtually any political opponent of the regime, or any "personal enemy" of any of the local eugenics courts [*Erbgesundheitsgericht*].)[14] Stoddard was certainly clear about who deserved credit for the intellectual justification of compulsory state sterilization of the "genetically unfit" in Germany. It was, he said, a "gift" from the American eugenics movement.[15] More frightening was what Charles Davenport wrote in 1925: "Our ancestors drove Baptists from Massachusetts Bay into Rhode Island but we have no place to drive the Jews to. Also they burned the witches but it seems to be against the mores to burn any considerable part of our population."[16] It is hardly surprising then that in 1932 Davenport was willing to defend Nazi extermination of the Jews.[17]

It should be emphasized that racial eugenics in the German intellectual community were put forward by members of the medical profession, for example Earnest Haeckel and Dr. Alfred

Ploetz. In 1905 Ploetz had written about "counter selection," by which he meant that everything from welfare to war might actually increase the numbers of "inferior" individuals. He frowned upon caring for the poor and the sick because it led to racial degeneration.[18] Also in that same year, the Society for Racial Hygiene was founded. By 1910, the German Government had recognized eugenics as a respectable part of German biomedical science.[19] The Kaiser Wilhelm Institute for Eugenics was founded in 1927, a time when Hitler was still busy starting fist fights in beer halls,[20] and less than a decade later the distinguished Dr. Ploetz was nominated for the Nobel Prize.

Between the world wars many German socialists identified eugenics with state planning and the rationalization of the means of production. Thus many, Left and Right, found the idea of a planned genetic future an attractive one.[21]

When the National Socialists came into power, biological science (their version of it anyway) was exceptionally important because it provided *scientific* support for their view of nature. The Nazis credited Dr. Ploetz with providing the theoretical foundations of their cause,[22] and Ploetz, embracing the Nazis, claimed that "National Socialism is the political expression of our biological knowledge."[23] This was shortened by Hitler's right-hand man, Rudolph Hess: "Nazism is applied biology."

While the Nazis pillaged biology and medicine for language in which they could express their eugenic and ideological goals, German intellectuals found that their political leaders were willing to support "scientific" experiments with unprecedented funding. Consequently the German medical community became avid in its support of Nazi policy and "eagerly embraced the racial ideal and racial state."[24] As Fritz Lenz wrote in 1933, "Whatever resistance the idea of racial hygiene may have encountered exists no longer. The German core [*Kern*] within the medical community has recognized the demands of German racial hygiene as its own; the medical profession has become the leading force in making these demands.[25]

Many of the doctors who embraced Nazi eugenics later found themselves convicted of crimes against humanity at the Nuremberg Trials. As noted in Chapter I, the prosecuting attorneys conceded that the Nazi leaders had broken no German civil laws, and so they appealed to moral law (natural law) in order to convict the indicted. The Court, and with it citizens of the civilized world, rejected the Nazi doctors' "scientific" dehumanization of Jews and Catholics, Slavs and Gypsies, and embraced instead the Judeo-Christian view of man as a *person* (not a mere individual), a person who is precious and possesses intrinsic value.[26]

One should be aware, however, that during the Nuremberg trials the defendants stressed that the models for German eugenics were programs that had been implemented in democratic nations.[27] As evidence, the defense attorneys filed a brief, "Race Protection Laws of Other Countries," compiled by the Information Service of the Racial-Political Office of the Reich Administration. It pointed a finger at many countries including our own:

### United States of America

Since 1907 sterilization laws have been passed in 29 states of the United States of America. Those affected by the law were primarily criminals, feeble-minded, insane, epileptics, alcoholic and narcotic addicts, as well as prostitutes. Although almost all states try to carry out sterilization on a voluntary basis the courts have more than once ordered compulsory sterilizations. (The brief then quotes the opinion of Justice Holmes in *Buck v. Bell.*)

### Denmark

Denmark was one of the first nations in Europe to pass a law permitting sterilization. . . . The vital interests of the community, as it says in the preamble, are to take precedence over the interests of the individual. The feeble-minded are sterilized.

### Finland

The draft of the Finnish law on the sterilization of persons with a hereditary disease goes back to 1929. The motion for the bill, which likewise provides for compulsory sterilization in specific cases, was passed in Parliament by a vote of 144 to 14.

### Norway

Norway also has a sterilization law. Efforts aim on the one hand at "securing a fertile breed" and on the other hand at "seeing that the nation is freed from parasites." Persons are sterilized who suffer from mental diseases or from imperfectly developed mental faculties and are therefore not capable of caring for themselves and their offspring by their own labor.

### Sweden

The Swedish Parliament has occupied itself with the question of sterilization since 1922 and in 1929 passed a law in this respect. The voluntariness which was expressed in this at first was annulled by an amendment in 1934. Compulsory sterilization therefore exists and is applied in cases of insanity. [28]

The defense also included as evidence the best-selling works of influential American eugenics writers Madison Grant (*Passing of the Great Race*) and Alexis Carrel (*Man the Unknown*). [29]

# 17

# THE NEW EUGENICS:
# GENETIC MANIPULATION AND CLONING

*In a society that came to view its members as just*
*so many cells or molecules to be manufactured or*
*rearranged at will, one wonders how easy it will be*
*to recall what all the shouting about "human rights"*
*was supposed to mean.*

Laurence Tribe

A FTER THE NAZI DEFEAT it was widely accepted that eugenics had been discredited. Nazi atrocities had shocked the
world, and now it became anathema to justify "population control" with blatantly racist or Social-Darwinist "ethics."
As a result the eugenics movement began to reinvent itself. *Annals*
*of Eugenics* became the *Annals of Human Genetics; Eugenics News*
was renamed the *Journal of Social Biology;* and the New Jersey
League for Sterilization was reincorporated as Birth Right, Inc.
Eugenicists now called themselves "population scientists," "human geneticists," "sociologists," and "family politicians." [1]

The climate of public opinion after Auschwitz forced the eugenics movement to do more than simply contrive new terminology. A New Age called for new issues, and the eugenics agenda became absorbed by the campaigns to prevent overpopulation and pollution, even if some of the policies recommended were little different than the ones so recently discredited in Nazi Germany.[2] In 1949 Garret Hardin, Ph.D. published a textbook entitled *Biology: Its Human Implications* wherein he called for "a painless weeding out before birth or a more painful and wasteful elimination of individuals [with low IQ] after birth."[3] In the 1960s the eugenics movement underwent a renascence of sorts with the founding of Voluntary Sterilization, Inc. and the People Pollute movement, the stardom of Paul Erlich and *The Population Bomb,* and the political campaign for Zero Population Growth. Ecological degradation was "blamed" on "the poor, the near poor, and the lower middle classes."[4] Dr. Curtis Wood, Jr., head of the Association for Voluntary Sterilization wrote in 1973 that "the welfare mess . . . cries out for solutions, one of which is fertility control [sterilization]."[5] By 1973 this sort of thinking had already tricked down into the elementary schools, as in the final stanza of "Overpopulation," a poem written by two sixth graders: "If we didn't have people,/We wouldn't have pollution./ Get rid of the people./That's the only solution."[6]

More recently, William Shockley, Nobel-prize winning co-inventor of the transistor, wondered: "Do our nobly intended welfare programs promote dysgenics—retrogressive evolution through the disproportionate reproduction of the genetically disadvantaged?" His answer, very much in the tradition of Margaret Sanger, is the proposition of "sterilization bonus plan" for various undesirables.[7]

The eugenics movement has managed to maintain its credibility with the population at large. Its constituents argue for a political agenda under the guise of liberal, "scientific," or "scholarly" programs. Legislators and judicial magistrates have sanctioned abortion, third party artificial insemination, *in vitro* fertilization, and fetal experimentation, and how infanticide and geronticide. Richard John Neuhaus noted recently that eugenics has returned "with a vengeance in the form of developments ranging from the adventurous to the ghoulish."[8]

Consider the following:

- Nobel-prize winners Francis Crick and James Watson, discoverers of DNA, argue that only after a rigorous examination should newborns be allowed to live. Crick also asserts that the time has come to question the assumption that people have a right to have children.[9]

- The American Civil Liberties Union argues in the courts that the fetus is only tissue. Some proponents of abortion go so far as to term the fetus "protoplasmic rubbish," or a "gobbet of meat." It should come as no surprise that the upshot is the legally commercial sale of fetuses, and in Europe, fetuses are used as protein ingredients in cosmetics.[10]

- *Insight* magazine reports in July 1988 that the National Institutes of Health (which hands out 85 percent of government funds for medical research) has been accused of failing to monitor whether fetuses have been declared dead when tissue has been taken.[11]

- An Indiana court allows parents to starve to death their handicapped baby.[12]

- A doctor suspected of killing 20 residents of a nursing home, pleads guilty to five of the murders, is convicted of three, and is sentenced only to pay a fine.[13]

- On May 26, 1988, a judge orders a young woman to remain on birth control for the rest of her childbearing years.[14]

• In the Summer, 1988 issue of *Daedalus*, biologist John Maynard Smith suggests that we give tax breaks and government allowances to university teachers who beget children and tax increases to others to discourage childbearing. (Can you imagine a world populated solely by the offspring of college professors?) Smith also points out that H.J. Muller and Julian Huxley postulate that married women should be permitted to have one child with their husband and then one by a donor of their choice. "In view of the sources from which it emanates" Smith writes, "if for no other reason, this suggestion merits careful examination."[15]

The new eugenicists have broken the genetic code and may soon enter the marketplace to peddle blueprints for the creation of the "perfect" human person. Biotechnologists are patenting their findings in the field of molecular biology, and it is anticipated that in the near future they will be marketing technological approaches to the reproduction process.[16]

Under the guise of curing diseases and preserving life, biotechnology corporations encourage people to sell them their blood, kidneys, sperm, eggs, embryos, fetal parts, human biochemicals, cells, and genes. Human body parts are now considered commodities.[17]

The door was opened for this practice in the fifties when the first commercial blood bank was incorporated in Kansas City, Missouri. Shocked that donating blood was no longer considered a civic duty, the local medical establishment created the non-profit Community Blood Bank in 1957. Accepting only volunteer donors, the bank signed exclusive contracts with most of Kansas City's hospitals and doctors. The commercial blood banks reacted by filing a restraint of trade complaint with the Federal Trade Commission (FTC). William Bennett, an FTC examiner, ruled, in September 1963, in favor of the plaintiffs. In his judgment the Community Blood Bank had conspired to restrain trade in the "commodity" of blood.

Outraged by the decision, the local community promptly appealed. The American Red Cross and the American Medical Association filed statements claiming that "trafficking in blood is unethical and immoral." [18] The medical community protested that human blood could not be treated as a commodity, because it is part of the human body's functioning process. The public furor induced Missouri Senator Edward Long to introduce federal legislation (which was not enacted) to exempt non-profit blood banks from antitrust legislation.

The FTC, meanwhile, supported the examiners decision: "There is a sufficient basis in the record for a factual conclusion that whole [human] blood is a 'biological product.' . . . As a result the commercial banks in this case when acquiring, processing, and supplying such blood to hospitals are engaged in the business of producing and selling a product. . . . The Commission clearly has jurisdiction to proceed against a conspiracy designed to have the effect of hindering the operation of such a business, and we so hold." [19]

In January 1969 the Eighth Circuit Court of Appeals reversed the FTC and exempted the non-profit corporation from FTC regulations. It did not, however, comment on the concept of human blood as a commodity.

Although hepatitis scares destroyed the commercial blood banks (their market share declined from eighty percent to one percent by the early seventies), a principle was established, namely that one could treat body parts as a commodities and put them on the auction block. Eventually the a federal court ruled (1976) that blood was just another organic product: "The rarity of the petitioner's blood made the processing and packaging of her blood plasma a profitable undertaking, just as it is profitable for other entrepreneurs to purchase hen's eggs, bee's honey, cow's milk or sheep's wool for processing and distribution. Although we recognize the traditional sanctity of the human body, we can find no reason to legally distinguish the sale of these raw products of nature from the sale of the petitioner's blood plasma."[20]

Today pharmaceutical companies dominate the blood market. Blood that contains lupus antibodies is valued anywhere from $50 to $200 a pint. Blood that does not contain a blood clotting factor fetches $600 a pint.[21]

The blood cases encouraged the commodification of other vital human organs. For instance, in 1983 Dr. Barry Jacobs opened the International Kidney Exchange (IKE). The kidney market was certainly lucrative, so much so that companies like IKE could sell one kidney for up to $160,000. Congress reacted by passing the 1984 National Organ Transplant Act (NOTA) which forbids the interstate sale of transplant organs. Penalties for violation can as much as five years in jail and a $50,000 fine. But this law contains loopholes; for instance, it does not forbid the sale of human organs for research, and it exempts "replenishable issues such as blood or sperm."[22]

Even with this U.S. prohibition the human organ commodity exchange flourishes internationally. In 1991 the World Health Organization (WHO) reported that sales of human organs in the Third World were reaching "alarming proportions": "Organ donation is undoubtedly a profoundly humane gesture, but its legislation and use without major restrictions involve one of the greatest risks man has ever run: that of giving a value to his body, a price to his life. Very many countries, be they poor or very rich, are . . . confronted with the increasing development of an organ market."[23]

As International Anti-Euthanasia Task Force Director Rita Marker pointed out, exploitation has become a major concern:

"Call it what we may, payment for organs is a bounty placed on the bodies of those whose families are least able to withstand financial pressure. . . . It will be the poor, the desperate and the disadvantaged whose loved ones will be worth more dead than alive."[24]

Serious ethical issues come into play concerning the overall integrity and value of the whole human person. Jack Kevorkian (and he is not the only one) has argued that since people have a right to control their bodies, they have a right to merchandise them in any way they wish. "Body parts are property," he has said, "The person owns them and has the absolute right over what will be done with them in every situation."[25] But against this view there are those who, like philosopher William May, refuse to believe that decisions concerning the body should be driven by the marketplace: "If I buy a Nobel Prize, I corrupt the meaning of the Nobel Prize. If I buy an exemption from the draft, which was permitted in the Civil War, I corrupt the meaning of citizenship. If I buy and sell children, I corrupt the meaning of parenthood. And if I sell myself, I corrupt the meaning of what it is to be human."[26]

In the mad race to procure organ parts and to gain commercial advantages, the definition of death has been manipulated. If brain activity ceases, even though artificial devices are keeping vital organ parts alive, a person is certified as dead. Center for Technology Assessment President Andrew Kimbrell, in his work *The Human Body Shop*, explained it this way:

> This new criterion for death provided three major advantages. First, in the hands of competent physicians, the diagnosis of irreversible loss of all brain function is said to be clinically practical and reliable. Second, the medical situation for such patients is viewed as hopeless; reportedly, these patients never regain consciousness, and suffer car-

diovascular collapse and cardiac arrest within hours, days, or in rare cases, weeks. Finally, as noted, these patients are an excellent source of organs. Under the brain death definition, surgeons could remove organs from patients who were "dead," but whose hearts were still beating and whose lungs, albeit artificially, were still breathing.[27]

The American Medical Association and a White House Commission have endorsed this definition. They approve of people in comas being treated as "living cadavers."

Yet even as this dramatic redefinition of death has been established, members of the medical profession have been pushing to change it again. They argue that death should mean the loss of "higher" brain functions (cerebral death), the loss of personal identity. In other words if a patient maintains "lower" brain functions and remains able to breathe on his own, he may still be declared dead and his organs harvested. Karen Anne Quinlan would have fallen under this category as would several thousand anencephalic babies born each year. Yale Medical Professor Dr. Robert Levine, who supports this broader definition of death, has suggested that an anencephalic baby "has more in common with a fish than a person."[28]

Reacting to this trend, for which there is no scientific proof, Charles Krauthammer wrote: "It is tragic that one cannot use the organs of a hopelessly doomed anencephalic to save the lives of other infants. But to kill one innocent for the sake of another is simple barbarism. Even just shortening a doomed life in order to lengthen another is a fateful start on that road. And it will not stop there. The anencephalic is the frontier case, and the frontier is always moving. Next comes the irreversibly comatose adult . . . then come the Alzheimer's patients. Why not bring some good out of their tragedy too?"[29]

The argument that a person dies but the body lives raises serious moral questions. Viewing these new definitions of death

by the materialist heirs of René Descartes, philosopher Hans
Jonas has observed:

> I see lurking behind the proposed definition of death, apart
> from its obvious pragmatic motivation, a curious revenant
> of the old soul-body dualism. Its new apparition is the
> dualism of the brain and the body . . . it holds that the
> true human person rests in (or is represented by) the brain,
> of which the rest of the body is a mere subservient tool.
> . . . [Yet] my identity is the identity of the whole organ-
> ism, even if the higher functions of personhood are seated
> in the brain. . . . Therefore, the body of the comatose, so
> long as—even with the help of art—it still breathes, pulses,
> and functions otherwise, must still be considered a residual
> continuance of the subject . . . and as such is entitled to
> some of the sacrosanctity accorded to such a subject by
> the laws of God and men. That sacrosanctity decrees that
> it must not be used as mere means.[30]

Taking advantage of the broader definitions of death, some
medical practitioners now plan abortions in order to procure infant
body parts. Arthur Caplan, M.D., a medical ethicist, has documented numerous cases of "harvesting," including that of a
diabetic woman who, after conceiving through artificial insemination, aborted the unborn child in order to obtain pancreas cells.
The cells were utilized in a transplant procedure to help improve
the woman's own health.[31]

Commercialization extends beyond the marketing of body
parts. Baby making is also a lucrative business. *In vitro* fertilization (IVF), gamete intrafallopian transfer (GIFT), Zygote
intrafallopian transfer (ZIFT), tubal embryo transfer (TET), partial zonal dissection (PZD), microsurgical epididymal sperm as-
piration (MESA), donor insemination (DI), egg donations, genetic
and non-genetic surrogate motherhood—these reproductive technology procedures are big money makers. Concerned about the

motivations of the various medical vendors, Dr. Janice Raymond has stated that she and other feminists are "angered that these technologies are being represented as safe, effective, and in a woman's best interest. . . . They are none of these things. . . . IVF clinics exist because they are immensely profitable. They aren't proliferating out of altruistic impulses for so-called desperate infertile women."[32]

Each year artificial insemination is practiced by over 11,000 American doctors and 170,000 of their female patients at a cost of $165 million. Sperm donations, at $200 per sample, create 65,000 babies each year.

But as in any growth industry, there are problems. Some donors of sperm and eggs have begun to assert parental rights. Studies show that sixty percent of paid donors have an interest in meeting their children when they turn eighteen.[33] Medical professionals fear that offspring of frequent donors in small communities may intermarry. Others are concerned that the traditional legal and moral meaning of parenthood is becoming extinct. Andrew Kimbrell is one who is alarmed: "Reprotech represents a disturbing alteration in our social and legal view of the human body and childbearing. It has thrown into doubt the very definition of parenthood. We no longer are sure what legally constitutes mother or father. We do not have a legal definition of human embryos—are they property or people?  And what of sperm and eggs—should they be viewed as commodities, no different from any other? . . . [O]ur society has produced few answers to these profound and unprecedented questions."[34]

Human eggs, in fact, are in great demand—laboratory-fertilized eggs can command a price of up to $12,000. Clinics maintain lists of women willing to provide eggs for compensation. Advertising that reads, "Healthy Women Wanted as Egg Donors. Help Infertile Couples. Confidentiality Ensured,"[35] has become commonplace.

Test tube babies (*in vitro*-fertilized embryos transferred to a mother) have become a routine procedures. Over 20,000 babies have been born since Louise Joy Brown was first created in July 1978. The freezing of human embryos for future transplants—a procedure familiar to cattle breeders—has also become common. Geneticists can screen embryos and determine genetic traits before implementation and destroy undesirable embryos.[36]

Frozen embryos raise serious questions: Do they have intrinsic value or are they a commodity that can be purchased, owned, or even patented? Some states have begun to grapple with these questions. Louisiana, for instance, has determined that an embryo is a "juridical person," who can inherit property if born. The state also forbids the destruction of frozen embryos. A Florida law states that "no person shall knowingly advertise or offer to purchase or sell, or purchase, sell or otherwise transfer, any human embryo for valuable consideration."[37]

Baby brokers, in an industry that is generating over $40 million in sales, hire women to bear children for others, with fees ranging from $30,000 to $45,000. Babies are treated like cattle futures, their lives are negotiated before conception. In a 1991 *Glamour* magazine article titled "Whose Child is This?", feminist leader Katha Pollitt voiced her concerns: "Surrogacy degrades women by devaluing pregnancy and childbirth; it degrades children by commercializing their creation; it degrades the poor by offering them a devil's bargain at bargain prices."[38]

A 1990 California Superior Court ruling in *Anna Johnson v. Mark and Crispina Calvert* changed that state's legal definition of motherhood. In the court's judgment the artificially inseminated woman, Anna Johnson, "is the gestational carrier of the child, a host in a sense."[39] Without citing any case law to support his finding, the judge ruled that technological advances had rendered irrelevant much of what has traditionally defined motherhood, most notably gestation and even birth itself. Because Mrs. Calvert's fertilized egg had been implanted in Ms. Johnson, the newborn was ruled to be the Calverts' biologically and, therefore, exclusively. Johnson was awarded no rights whatsoever, not even visitation rights. Reacting to this decision, which allows motherhood to be sold as a commercial product, surrogacy expert Dr. Michelle Harrison stated: "Home is where the womb is. For the slowly evolving fetus, home is the place filled with the mother's warmth and cushioned by her fluids. . . . In the last months, the fetus hears her singing, her talking, her crying. At birth the newborn shows preference for her voice above all others. . . . The donor of the egg did not experience the nine month long intimate, dynamic and life giving process that went on for every second of gestation. In the fullest biologic sense, the donor of the egg has not mothered that baby." [40]

Although the marketing and engineering of the human person has proven to be financially profitable, this is not the sole criterion that drives the new eugenicists. Through continued technological advancements, they believe they will fulfill Francis Galton's dream of breeding a "flawless person."

Pre-implant genetic screening is now being utilized by parents to determine if their prospective child's genetic traits are desirable. Microscopic examination of cells, from amniocentesis and chronic villus sampling, can now detect over 200 genetic disorders. [41] Parents dissatisfied with the genetic analysis often turn to abortion. Twenty years ago, when a woman gave birth to an impaired child, some people would say, "Oh, what a heartbreak that she has to live with that situation." "But you'd look for reasons for it," says writer Deborah Batterman, who had amniocentesis during a recent pregnancy. "You'd say, 'I have to examine my life and see why this happened and what I can do to make the best of the situation.' That's the martyrdom indoctrination. We don't have to do that now. We're not raised to be martyrs."[42]

Genetic screening is not just employed to detect diseases; it is also used for sex selection. A 1988 survey revealed that twenty percent of geneticists approve of this form of prenatal "diagnosis." Although doctors are not required to record reasons for abortions, medical professionals have confirmed that sex-related abortions are on the rise. In July 1987 a National Institutes of Health ethics specialist, Dr. John Fletcher, told *New York* magazine: "The difference between having a baby twenty years ago and having a baby today is that twenty years ago, people were brought up to accept what random fate sent them. And, if you were religious, you were trained to accept your child as a gift of God and make sacrifices. That's all changing."[43]

The Federal government is currently funding the $3-billion Human Genome Project, the intention of which is to decipher all 100,000 human genes. The National Institutes of Health awarded a $600-thousand grant to study genes that determine IQ. These and other research projects anticipate the ultimate materialist revelation: that all physical and behavioral traits can be explained genetically.[44]

Parents will soon have the power to choose an embryo that will meet their standards. They will be able to choose to abort a child because tests indicate that he will be short, or bald, or fat, or of average intelligence, or simply because they don not like the color of his eyes. Reflecting on this trend, University of Chicago professor of obstetrics and gynecology, Dr. Mary Mahowald concludes that "It is eugenics. We don't give it that name, but we foster the concept nevertheless. It has intensified over the last decade because of the two child family, the availability of abortion and the techniques we have for pre-natal, even pre-pregnancy, diagnosis. All those together contribute to the notion that people not only ought to be able to determine when to have children and how many to have, but also just what kind of children to have.[45]

"Wrongful-life" suits are now filed in the courts. In *Turpin v. Sorline* (1982), *Procamile v. Cello* (1984), and *Habeson v. Parke-Davis* (1983) juries decided that children born with certain maladies have legitimate tort claims against their obstetricians. In these cases the parents, representing the child, sue on the grounds that since the obstetrician did not properly inform the parents during pregnancy that their newborn might not be in perfect health, they (the parents) were denied the opportunity to choose abortion and are consequently entitled to compensation.[46] In the September 15, 1997 edition of *The Weekly Standard* David Tell points out that "the alleged tort, in such a case, is life per se. The plaintiffs' newborn argues, essentially, that he has been denied the right to be aborted—that his undesired life is less valuable to him than non-existence would be."

While parents are concocting their "flawless baby" checklists, the biotechnology industry, in the name of productivity, utility, and profit, is rapidly advancing plans to control the blueprints of life.[47] On February 22, 1997 Dr. Ian Wilmut of Scotland's Roslin Institute revealed to the world that he had successfully cloned an adult Dorset sheep. "What makes Dolly [the cloned sheep] different," reported the Institute's director, "is that she was not cloned from sex cells, but from mature mammal cells with no reproductive function."[48]

Dr. Wilmot had procured a cell from a Finn Dorset ewe and starved the cell in order to terminate the dividing process. The cell was then placed next to a Scottish Blackface ewe's unfertilized nucleus, essentially a barren cell. Electric pulses fused the cells causing fertilization and stimulating cell division. This new cell, now an embryo, was implanted in a Blackface ewe's uterus, and she gave birth to a Finn Dorset lamb genetically identical to the original donor. [49]

The international reaction was capsulized in the title of *Time* magazine's March 10, 1997 cover story: "Will there be another you?" Philosophers, theologians, psychologists, sociologists, and people on the street pondered the potential impact of Dr. Wilmut's creation. The fundamental question on everyone's minds was now not if a man can be cloned, but if a man should be cloned.

The subject of cloning is not new; scientists began researching the possibility of asexual reproduction in humans (parthenogenesis) in the late nineteenth century. The intent has been to artificially create offspring by utilizing a cell procured from a living being. "The procedure," according to Dr. Joseph Fletcher, "is called asexual or nonsexual because the baby produced from it is not the result of a combination of male and female gene cells. The body cell may itself be the result of sexual reproduction but now it is alone the starting point of a new individual, it does not combine together with any other sex cell." [50]

As early a 1896 the German embryologist Oskar Hertwig created sea urchins asexually, and overtime the process was so improved that by the 1950's scientists had identified over 370 spermless chemical and biological techniques for sea urchin reproduction. [51] The theory behind this phenomena was expressed by Dr. Ernest Messenger in his 1931 book *Evolution and Theology*: "In principle, every cell in the ordinary organism contains the virtuality of the species and the race."

Cloning became a reality in 1952 when researchers at the Philadelphia Institute for Cancer took the nuclei of a frog cell and substituted it in the egg of another frog. The result was a frog genetically identical to the donor: a clone. From that point forward the challenge was learning how to apply the finding to mammals.

By 1988, thanks to grants from W. R. Grace and Company to the University of Wisconsin, the nuclear transfer technique used with frogs was utilized successfully to clone cattle embryos. The genetic structure of award winning steers could now be reproduced by having their cells inserted into eggs and placed in cows for gestation. The *New York Times* applauded this patented technique because it added "factory-like efficiency to animal reproduction."[52]

There were, however, side effects to the cloning of embryos. Twenty-one percent of the calves born were abnormally large, and five percent were giants. The mishaps caused one ethicist, Dr. Andrew Linzey, to make this comment: "These are attempts to create animal machines that yield more milk, or tastier meat or bigger profits. To manipulate their lives for these ends is morally grotesque. When the end result is freak animals, inevitably there will be additional suffering."[53]

Dr. Robert Stillman and his colleague Dr. Jerry Hill received the "general program prize" at the 1993 American Fertility Society annual meeting for their research paper describing a formula that transformed seventeen human embryos into forty-eight embryos.[54] When the media broke the story, government and church leaders, concerned about the experiment's social implications, voiced their outrage and horror. But there were those who reacted favorably, because they detected moneymaking opportunities. Selling "super embryos" could reap big profits for the fertility industry.

The announcement on March 2, 1997 that scientists at the Oregon Regional Primate Center had cloned two rhesus monkeys added to the frenzy.

Grasping the frightening consequences of human cloning, the Vatican called for a ban on the procedure. John Cardinal O'Connor, Archbishop of New York, explained that "from the perspective of the Church, the whole concept of human cloning is morally repugnant. . . . The Church would consider any experiments with human cloning immoral. It's our conviction that all procreation should come from natural relationships between a man and woman." [55] Foundation on Economic Trends president, Jeremy Rifkin referred to cloning "as a horrendous crime." He continued: "You're putting a human into a generic straitjacket. For the first time we've taken the principles of industrial design—quality control, predictability—and applied them to a human being." [56]

With opinion polls showing that ninety-three percent of Americans oppose human cloning, Congress passed legislation forbidding U.S. financing for human embryo research. President Clinton, citing loopholes in the law, ordered a further ban on the spending of federal dollars. In his March 4, 1997 statement the President said, "I believe we must . . . resist the temptation to replicate ourselves." He also asked for a voluntary moratorium on private funding for cloning research until his National Bioethics Advisory Commission released their study on the legal and ethical implications of human cloning. [57]

On May 18, 1997 the *New York Times* reported on the Commissions progress:

> The ethical arguments are thorny. At a committee meeting last month, Dr. Bernard Lo, the director of the Program in Medical Ethics at the University of California in San Francisco, said the group was "grappling with what it is about cloning that has raised such strong emotions." Although some ethicists and religious leaders have argued that cloning would violate human dignity or would endlessly complicate family relationships or that it would be the ultimate hubris, Dr. Lo said, "the problem the group confronted was that there was no one compelling reason why cloning should

be banned." Nonetheless, he noted, "There was no easy way to dismiss the religious, almost mystical argument that it was deeply objectionable, an affront to human dignity."[58]

The Bioethics Commission presented to President Clinton its recommendations in June. Calling human cloning at this time "morally unacceptable," the commission endorsed the extension of the moratorium on the use of federal money to support cloning of humans. The commission also encouraged passage of congressional legislation that would prohibit private research groups from receiving government funding for three to five years.[59]

The press release announcing the findings was deceptive, however. The commission's legislative recommendations did not call for a ban on the cloning of animal and human genes and cell lines, and by implication, other body parts.[60] The commissioners did ask for a suspension of "somatic" cell cloning of children, but, most importantly, their call for a temporary ban was *not* based on ethics, but rather on safety.[61] Senator Christopher Bond voiced his disappointment: "I had hoped the Federal ethics commission would not be afraid to make a strong moral statement. But when it came to the tough questions they punted and now it will be up to Congress and state legislatures to resolve those issues." The United States Catholic Conference took a similar position: "[T]hough labeled a prohibition . . . [the commission's report] contains everything a researcher could want to ensure a future for human cloning—and to allow much destructive research on human embryos in the meantime."[62] The commission took a dive, refusing to come to terms with the fundamental moral and ethical issues that revolve around the cloning issue. The commissioners failed to consider if cloning threatened the dignity of the human person.

18

# Does Human Cloning Violate the Dignity of the Person?

*Throughout our nation's history, the Judeo-Christian tradition has respected the divine design of life-giving love. In the process of cloning, this personal, unitive, two-in-one-flesh dimension of life-giving marital love is rejected and replaced by technological replication. Begetting is the continuation of creation; manufacturing is proper to and productive of things, not persons.*

John Cardinal O'Connor

T
HE WORLD is a machine in which there is nothing at all to consider but the shapes and movements of its particles."[1] This statement of Descartes has had an incredible influence on modern thought. By strictly reducing science to various versions of materialism, physicists, political scientists, economists, and psychologists can (and, by the logic of their

arguments, *must*) treat man and beast alike, as machines differing only in degree of complexity. Thanks to Cartesian reductionism, psychology especially is no longer the study of man as a spiritual being possessing body and soul, but is merely biology, the study of cells. Biology is then reduced to the study of organic chemistry. Chemistry is reduced to the study of physics in which, finally, man is an organism in which quanta swirl aimlessly.[2]

Based on this view, man the cell, man the chemical compound, is no more responsible for his creative accomplishments in music, art, literature, scholarship, science, and invention, than the warthog is responsible for his warts. Moral choice can be nothing more than the tropism of an automaton conditioned by various genetic, social, and historical contingencies.[3]

The pseudo-humanism of the mechanists believes: that there is no difference between man and brute; that empirical evidence alone constitutes the knowledge of the phenomenon called man; that there is no "objective" existence of mind, consciousness and the soul; that human freedom is illusion; that there is no human nature which is not malleable to techniques of design, development and control.[4] These notions permit biologist Stephen Jay Gould to assert that our "brains are enormously complex computers,"[5] and they permit psychologist B.F. Skinner to categorically state that man is a "machine in the sense that he is a complex system behaving in lawful ways." Carl Sagan declared that "the oak tree and man are the same. . . . At the heart of life, we are identical to a tree," and Dr. Malcolm Watts postulated that it will "become necessary and acceptable to place relative rather than absolute values on such *things* as human lives, the use of scarce resources, and the various elements which are to make up the quality of life or of living which is to be sought." And Dr. Watts has proposed "a new ethic in a rational development of what is almost certain to be a biologically oriented world society."[6] Thus has man become a commodity.

Analyzing the progress of the mechanistic view of man during the past century, Andrew Kimbrell made these observations:

> The doctrine of mechanism was significantly refined as Western civilization entered into the industrial age. As more complex and sophisticated machines were developed, the machine image of the body also evolved. By the twentieth century, mechanism's proponents had set out to remake the body in the likeness of the modern motor with its greatest attribute—efficiency. These attempts were to have extraordinary consequences for human work and the human body shop. They were also to lead directly to perhaps the most pernicious practice of the twentieth century: eugenics.[7]

Since man is merely a machine and there are no moral standards, contemporary eugenicists can easily argue for human cloning as a means to end "reproductive roulette." Hence Dr. Joseph Fletcher, the medical ethics professor who authored *Situation Ethics*, can say: "Good reasons in general for cloning are that it avoids genetic diseases, bypasses sterility, predetermines an individual's gender, and preserves family likenesses. It wastes time to argue over whether we should do it or not; the real moral question is when and why."[8]

There is, of course, a perspective that rejects the mechanistic calculations of the eugenicists. Its proponents hold that all things come from God, especially those attributes that make man Man: his reason, his imagination, his creativity, and his moral and aesthetic dimensions. They argue that the spiritual entity, the mind, with its power of reflection makes man substantially different from the beast. For them the existence of the mind forcefully undermines the foundations of the mechanists. Because with a mind man can have choice, he can have will, and he can have reason and values; with a mind man is different from the rat.

This school of thought holds: that the existence of God is the central focal point of human existence and history; that spirit and flesh are unified in the mind of man; that the soul, reason, imagination, will, and man's moral and aesthetic dimensions, although imperfect, are all gifts from God and exemplify the Divine spark in man; that as a consequence, *every* individual *matters* and is a universe in himself; that we must reject the neo-gnosticism of current scientific thinking, including the fallacies of positivism, the claims of empirical and physical reductionism, and the blatant de-humanizing aspects of behavioral determinism as "the primal twilight of man;" that we much achieve the total rejection of the concept of a malleable human nature whereby individuals, the political and social structures, culture, and human evolution are to be re-cast to fit the specifications of elitist, technocractic planners through behavioral *control* and genetic *manipulation*, and embrace the belief that human nature is *constant*; and finally, that the most important things in life must be left to the to the decisions of the *common man*.

As part one of this book described, man, made in the image and likeness of God, is something special; he has intrinsic value. By his nature, he is a social being who possesses inalienable rights. The state, having received its power from God through the natural law, has the duty to protect man's God-given right to life and to uphold his dignity by maintaining and enhancing the common good.

As a social being, man is endowed by God with an appetite and inclination for social life and naturally forms the basic unit of society, the family. The bond of marriage is a natural institution. Nature intends the continuation of the human race, because it has given human beings the faculty and instinct for reproduction.

Based on these tenets, human cloning technology would be an affront to the person's human dignity and a violation of the common good.

Human cloning violates man's personhood; it denies the sanctity of the person. A cloned child would be an object, a product of scientific engineering, and not, legally if not actually, a gift of God. As John Cardinal O'Connor points out: "There remains a profound ethical difference between 'having a child' and 'making a child.'"[9] Instead of being accepted as a person, the cloned child might be treated as "thing" or a "chattel." To think that such a prospect is impossible is to ignore the lessons of history.

Human cloning technology could assault the basic fabric of society—the common man and his family. The bonds of marriage could be rendered meaningless. Testifying before a United States Senate Committee hearing on human cloning in June of 1997, John S. Bonnici, S.T.D., stated "Human cloning technology seeks to replicate the person without one individual having to enter into relation with another. This dynamic contradicts man as a social being. Furthermore, without relations with others each individual cloned is suddenly unable to truly live and develop one's gifts. This unnatural act comes at great cost. The potential for a well-ordered and prosperous society is no longer anticipated. The family—the original cell of social life—is compromised by the unnatural objective associated with human cloning."[10]

This potential biotechnology advancement raises an inexhaustible number of legal and ethical questions. Thousands of experiments in embryo cloning have resulted in mishaps. If experimentation in human cloning produces misfits or freaks (as happened with cattle cloning), how will they be treated? If they are considered commercial products, as "things," will they be tossed into a shredding machine if they are less than perfect? Will they be recognized as persons entitled to rights, or will they be treated as animals or slaves or robots without any legal protections? Who will be responsible for misfits or unwanted "products"? Will the biotechnologist be responsible for the well-being of his creation? Will the "product" be treated as a child or a pet? Will the "product" be entitled to a room of its own or a cage?

Human cloning technology could be easily abused. Nazi doctors, for instance tried to link science with the ideals of social Darwinism, racial hygiene, and Aryan supremacy.[11] They envisioned "breeding a new human type," "creating a new man."[12] And experiments to achieve that vision began in the death camps.

It is conceivable that a government, dominated by legal positivists and utilitarians could easily abuse cloning technology and create an army of super-soldiers to enslave its citizenry. No, cloning technology is not compatible with a society that cherishes the dignity of the human person.

The Dean of Humanities at the Massachusetts Institute of Technology told a meeting in 1949: "We must now recognize our approaching scientific ability to control men's thoughts with precision."[13] It so happened that Winston Churchill was an honored guest on the occasion, and he remarked in reply that he would "be very content to be dead before that happens."[14] Churchill, who recognized the inherent evil of twentieth century ideologies, surely agreed with G. K. Chesterton's insight that all men matter. "You matter. I matter. It's the hardest thing in theology to believe." Yes, even the most degraded and unknown soul that ever walked the face of the earth is still not the dross of history, although that is what the eugenicists would have us believe, but is a universe in himself; a person of ultimate importance.

PART IV

# THE DEATH PENALTY

# 19

# MERELY VENGEANCE
# OR MORALLY JUSTIFIABLE?

*[T]he death penalty has symbolic significance. Those who favor it believe that the major remedy for crime is punishment. Those who do not, in the main, believe that the remedy is anything but punishment. They look at the causes of crime and conflate them with compulsions, or with excuses, and refuse to blame. The majority of the people are less sophisticated, but perhaps they have better judgment. They believe that everyone who can understand the nature and effects of his acts is responsible for them, and should be blamed and punished, if he could know that what he did was wrong. Human beings are human because they can be held responsible, as animals cannot be. In that Kantian sense the death penalty is a symbolic affirmation of the humanity of both victim and murderer.*

Ernest Van Den Haag

F OR DECADES the death penalty issue has haunted the halls
of Congress and the nation's fifty state legislatures. Ordinary citizens have anguished over the issue at their kitchen
tables and in the nation's jury deliberation rooms. Political ca-
reers have been made and destroyed, elections won and lost de-
pending on a candidate's position on capital punishment.

New York governors Hugh Carey and Mario Cuomo annu-
ally vetoed death-penalty legislation that for twenty years was
approved by the state legislature. Political analysts agree that
George Pataki's 1994 upset of three-term incumbent Cuomo was
due in large part to Pataki's pledge to approve the reinstatement
of the death penalty the moment the 1995 bill arrived on his
desk. And he did.

Today thirty-seven states have codified offenses punishable
by death. A vast majority permit juries to impose capital punishment for first degree murder while others, such as New Hamp-
shire, limit execution to convicted murderers of police officers
and kidnapping victims.[1]

In the 1990s the most aggressive death penalty statutes have
been instituted on the federal level, and there are now more than
three dozen federal capital offenses, including drug trafficking,
murder of an American citizen abroad, torture outside the United
States, hostage taking or kidnapping that result in death, aircraft
hijacking or carjacking that result in death, assassination of the
President, Vice President, members of Congress, cabinet members, and members of the Supreme Court, espionage, treason,
train sabotage that results in deaths, and murder at a U.S. international airport.[2]

Over the past decade public-opinion surveys have consistently confirmed that approximately seventy percent of Americans support the death penalty for those convicted of serious
crimes. Nonetheless, the debate rages on.[3]

A cover story of *U.S. News and World Report* about convicted Oklahoma City bomber Timothy McVeigh was entitled: "The Place for Vengeance—Many grieving families seek comfort and closure in the execution of the murderer. Do they find it?"

"The sooner [McVeigh] meets his maker, the sooner justice will be served," said Darlene Welch, whose four-year-old niece, Ashley, was killed in the blast.[4] "He will get what he deserves in the afterlife, where he will meet Hitler and Jeffrey Dahmer," said Ernie Ross, who suffered serious injuries from the blast while working across the street.[5] Another survivor hoped that McVeigh "would have one leg amputated and then be suspended over sharpened, growing bamboo shoots that would pierce his body."[6]

While the families of the Oklahoma City victims called for vengeance, the *New York Times* editorial page declared: "Neither our own recoil at the heinous deed nor our sympathy for the blast's many victims alters the basic concern that has long driven this page's opposition to the death penalty. We see capital punishment, which is state-sponsored killing, as morally wrong and against the Constitution's ban on cruel and unusual punishment. Public safety concerns and the understandable desire for revenge can be met even in this case by giving Mr. McVeigh a life sentence without parole."[7]

When the McVeigh jury ordered his execution, editorial responses were polarized:

- *The Boston Globe*: The calculated taking of life by the state brutalizes society and only perpetuates the evil. It does nothing to restore the lives of the victims. Human rights—the most fundamental of which is the right to life—cannot be ranked on a hierarchical scale. Capital punishment places the meaning of certain lives above the life of others which devalues them—us—all.[8]

• *The Norwich* [Connecticut]*Bulletin*: McVeigh irretrievably broke the basic aspect of the social contract: that citizens be allowed to live in peace. For him to have wiled away his life in jail would have diminished the lives he stole. Justice is served.[9]

Because the opponents in the death-penalty debate often argue at a fever pitch, we may forget that both sides base their positions on certain philosophical, theological, and ideological suppositions that give rise to their respective concepts of justice, punishment, and the nature of man and his role in the community. Judicial decisions, legislative action, and even editorial opinions are also influenced by these principles.

Is capital punishment merely vengeance? If so, is it morally justifiable? Is it a deterrent? Is the death penalty cruel and unusual punishment? Or is it the inherent moral right of the state? These are the issues that will be addressed in this section.

# 20

# THE ORIGINS OF
# THE DEATH PENALTY

*He that sheds the blood of a man, for that man
his blood shall be shed; for in the image of God
has God made man.*

Genesis 9:6

THE SENTENCE OF DEATH for horrendous crimes has a long
history in western civilization. Its justification is based
on biblical and natural law traditions and the concept
of justice that stems from those foundations.

The basis of political thought from Aristotle to Aquinas is
that man is a social being, that man "is by nature a political
animal." By his rational nature, a person forms a family and a
community based upon cooperation for their common good. The
formation of the state is natural and essential for the maintenance of the good for all members of the society.

Through the natural law God wills the existence of the state—the civil power without which human beings cannot achieve their destiny in this life. The primary purpose of the civil power is to protect and preserve the life and property of the citizenry. This cannot be achieved unless order is preserved in the community. Thus the state has the right of instituting civil laws to supplement the natural laws.

Laws exist to be enforced and enforcement is meaningless without the right to punish. Punishment exercised by a duly recognized government has three functions: retribution, to force the criminal to pay for his crime and vindicate the rights of the offended; correction, to rehabilitate the offender; and deterrence, to forestall similar crimes by warning the entire community.

The state has the right to exact retribution from a citizen who breaches public order, but it also has the responsibility to attempt to reform the criminal. In addition, the state must try to convince the citizenry that committing an injustice is wrong.

Retributive punishment seeks to secure justice, not revenge. Revenge is merely an emotional action.

The death penalty is viewed as retribution insofar as it attempts to adequately punish the most serious criminals—murderers and traitors, for instance. Aquinas wrote that in order to promote the common good "the state executes pestiferous men justly and sinlessly in order that the peace of the state may not be disrupted."[1] And we tend to agree when the murderer in question is particularly "pestiferous." When, for instance, the Israelis' executed Adolf Eichmann, very few had second thoughts about the execution of one who was one of the masterminds of genocide. And Eichmann was not executed because his punishment was likely to deter others but simply because he deserved to die. Common sense dictates that by his criminal actions he had forfeited his right to live.

In the Judeo-Christian tradition, one can also appeal to biblical texts concerning the state's right to inflict the death penalty: "And he that killeth any man shall surely be put to death." (Leviticus 24:17). "Who so ever sheddeth man's blood by man shall his blood be shed." (Genesis 9:6) "You shall not accept payment for the life of a homicide guilty of a capital offence; he must be put to death." (Numbers 35:31)

According to *The Code of Maimonidies,* Jewish law states that while it is proper to mourn for deceased relatives, no mourning should be observed "for those who have been condemned to death by the court."[2]

Christ also recognized the authority of the state and accepted its verdict of death. When Jesus stood condemned before him, and Pilate said: "I have the authority to release you, and I have authority to crucify you . . ." Jesus replied: "You would have no authority at all over me if it had not been granted you from above." (John 18: 9-11) St. Paul reiterated that because the state's power to punish is God-given, such punishment must be just: "I am now standing before the Emperor's tribunal, and that is where I am to be tried. . . . If I am guilty of any capital crime, I do not ask to escape the death penalty; but if there is no substance in the charges which these men bring against me, it is not open to anyone to hand me over as sop to them. I appeal to Caesar." (Acts 25:11-12)

In line with Christ's teaching, St. Paul sanctioned the state as an agent of "God's wrath" and recognized its power to judge, to punish bad actions and to reward good ones:

> Every person must submit to the supreme authorities. There is no authority but by act of God, and the existing authorities are instituted by him; consequently anyone who rebels against authority is resisting a divine institution, and those who so resist have themselves to thank for the punishment they will

receive, For government, a terror to crime, has no terrors
for good behavior. You wish to have no fear of the authorities? Then continue to do right and you will have their
approval, for they are God's agents working for your good.
But if you are doing wrong, then you will have cause to
fear them; *it is not for nothing that they hold the power of the
sword, for they are god's agents of punishment, for retribution on
the offender.* That is why you are obliged to submit. It is an
obligation imposed not merely by fear of retribution but
by conscience. That is also why you pay taxes. The authorities
are in God's service and to these duties they devote their
energies. (Romans 13:1-6)

DURING THE SUMMER OF 1994, while waiting to appear as a guest
on a Syracuse radio talk show, I heard Governor Mario Cuomo
tell the show's host that while he held strongly to his Catholic
faith, he would never impose his beliefs on anyone. And then
he added: "Incidentally, I feel that way about the death penalty
as apparently do most Catholics. Because as you know the Catholic
Church teaches the death penalty is wrong, but that does not
stop Catholic legislators from your area or elsewhere from vot-
ing for it."

Shortly thereafter, Joe Conason, executive editor of *The New
York Observer,* critiqued gubernatorial candidate George Pataki's
appearance on a television interview: "I have never heard a
Catholic candidate yet say 'I have to be against the death penalty.' George Pataki sat there and said 'I have great respect for
the teaching of my church.' Except all he did was talk about
the death penalty, which the church opposes! When is he going to be consistent with that teaching of the church?"

Although these prominent liberals turn, when expediency
moves them, to the Catholic Church to support their positions,

on this subject, they are wrong. The Roman Catholic Church has always acknowledged the state's power to impose the death penalty.

Pope John Paul II has often appealed for compassion and clemency for condemned murderers, and American Catholic bishops have stated that the death penalty should not be imposed in the United States. Nevertheless, the pope has never used his office to condemn capital punishment per se, and the bishops, whether taken singularly or collectively, have no authority under civil or canon law to urge the imposition of or attempt to block the application of the death penalty. That authority is vested solely in the civil power, and is consigned to the state by virtue of the natural law.

Representatives of the Catholic Church are free at all times to express their personal opinions that other forms of punishment are sufficient to ensure the proper order on which the well being of the community at large depends. But the determination that the imposition of the death penalty is necessary belongs exclusively to the state. The Church recognizes this power and understands that its source is Divine. What the Church does not confer, the Church cannot take away. Even the American bishops' statement opposing the use of the death penalty clearly admits that "the state *has the right* to take the life of a person guilty of a serious crime."[3] [Italics added.] John Cardinal O'Connor, who is personally opposed to the death penalty, stated from the pulpit in St. Patrick's Cathedral, that "formal official Church teaching does not deny the right of the state to exercise the death penalty under certain, narrowly defined conditions. It is a matter of judgment."[4]

Here is what the newly issued *Catechism of the Roman Catholic Church* states: "Preserving the common good of society requires rendering the aggressor unable to inflict harm. For this reason the traditional teaching of the Church has acknowledged as well-founded the right and duty of legitimate public authority to punish malefactors by means of penalties commensurate with the gravity of the crime, not excluding, in cases of extreme gravity, the death penalty.[5]

In reaffirming its traditional teaching on capital punishment, the Church in no way requires the state to exercise the death penalty, which is simply a prudential option. The *Catechism* also urges mercy, but again recognizes that leniency is granted at the discretion of the state.

The *Catechism* is promulgated by supreme authority to be used throughout the whole church. Its authority derives from the pope, who is, according to Section 331 of the Canon Law, the supreme authority, and by virtue of his office enjoys "full, immediate and universal ordinary power in the Church which he can always freely exercise."[6] The *Catechism*, therefore, supersedes any other Catholic voice, however esteemed.

The Catholic Church also dismisses the "seamless garment" proposition which argues that if one is opposed to abortion, one must be against the killing of any life for any reason. This argument has been widely used by those who wish to blur the obvious distinction between abortion and capital punishment in order to strengthen opposition to the latter. Abortion, according to the church, is wrong because it destroys innocent human life. Capital punishment is permissible because the first duty of the state is to maintain order for the common good. To meet this end, it is permissible for the state to kill those who are found guilty of grievous offenses in times of peace and war. Anyone who cannot see the obvious distinction between taking the life of an innocent child and taking the life of a man guilty of serial murder cannot be taken seriously as a teacher of morals or as a political leader.

# 21

# OPPOSITION TO THE DEATH PENALTY: IDEOLOGICAL ORIGINS

> [T]he Marquis of Beccaria, moved by sympathetic
> sentimentality and an affection of humanitarianism,
> has asserted that all capital punishment is illegiti-
> mate. He argues that it could not be in the origi-
> nal civil contract, inasmuch as this would imply that
> everyone of the people has agreed to forfeit his life
> if he murders another (of the people); but such an
> agreement would be impossible, for no one can dis-
> pose of his own life.
>
> Immanuel Kant

AN EARLY TRACT opposing the death penalty, *On Crimes and Punishment*, was authored by Cesare Beccaria in 1764. His work applied the "new" physical and moral principles of the Age of Enlightenment. Influenced by Thomas Hobbes, Beccaria advanced his mentor's position that the new laws *excluded* both the Bible and the natural law, and that "the basis of human morality" is "the expressed tacit compacts of men."[1]

Hobbes, the father of modern statism and the first advocate of totalitarianism, postulated that the state is founded on man's right of self-preservation. He assumed that man is naturally in a state of complete liberty and is driven only by his passions and desires. "Every man for his part, calleth that which pleaseth and is delightful to himself, good; and that evil which displeaseth him. . . . And as we call good and evil the things that please and displease; so we call goodness and badness the qualities or powers whereby they do it."[2] Since each person is completely free to do as he wishes, each person is free to violate the freedom of other people. Hobbesian man is vain, contentious, revengeful, and self-seeking; his primitive anti-social "state of nature" leads man to a state of constant warfare and hostility. Because of a constant state of material fear, man is "a wolf to man." There is a "war of all against all" with "no justice because there is no law."[3] The natural law and the traditional view that "man is by nature a political animal" is rejected by Hobbes. The natural origin of the state is replaced with the contractual theory of the state.

Men, due to their impulse for self-preservation and the realization of the incompatibility of competing interests, come together by compact (the general will) and cede their natural freedom to a sovereign who makes and enforces law. According to Hobbes:

> The only way to erect such a common power . . . is to confer all their power and strength on one man, or upon one assembly of men, that may reduce all their wills by plurality of voices, unto one will; which is as much as to say, to appoint one man or assembly of men, to bear their per-

son; and everyone to own, and acknowledge himself to be the author of whatsoever he that so beareth their person, shall act or cause to be acted, in those things which concern the common peace and safety; and therein to submit their wills, every one to his will, and their judgments to his judgment. This is more than consent, or concord: it is a real unity of them all, in one and the same person, made by covenant of every man with every man, *I authorize and give up my right of governing myself, to this man or to this assembly of men, on this condition, that thou give up the right to him, and authorize all his actions in like manner.* This done, the multitude so united in one person, is called a commonwealth. . . . For by this authority, given him by every particular man in the commonwealth, he hath the use of so much power and strength conferred on him, *that by terror thereof,* he is enabled to perform the wills of them all, to peace at home, and mutual aid against their enemies abroad.[4] [Italics added.]

The populace irrevocably conveys its liberties to the state; its private judgments to an absolute sovereign. To control the passions and judgments of the populace the sovereign has terrifying power, since, in Hobbes' words, "covenants without the sword are but words and of no strength to secure a man."[5] The head of state is absolute, indivisible, and inalienable, and he defines the "natural law" as he sees fit, which means natural law is ignored.

Accepting Hobbes' compact theory, Beccaria contends that leaders do not have the right "to slaughter their fellow beings." In his view, no man who joined the commonwealth "can have wished to leave to other men the choice of killing him."[6]

If humanity's sole motivation is self-preservation, no man would give the implied power to the sovereign to take away his life. Reflecting on the implications of Beccaria's thesis, twentieth-century political philosopher Leo Strauss observed that if the commonwealth is founded to preserve the life of its subjects then the state can "hardly demand from the individual that he resign that right . . . by submitting to capital punishment."[7]

Although Beccaria's arguments are quoted to this day, it should be noted that the great ideologues of the Enlightenment period who subscribed to the compact theory—Hobbes, Locke, and Rousseau—rejected his position on the death penalty. Locke stated: "Political power, then, I take to be a right of making laws with penalties of death and consequently all less penalties."[8] Rousseau concluded that if the reason for the compact between man and commonwealth is for the preservation of both parties, then he "who wills this end wills the means also."[9]

Beccaria's influence went far beyond the compact theory analysis. Like most thinkers in his time, he articulated the position that punishment should not be based on the "barbaric" notion of retribution but instead based on a reformed penal system that recognized crime as a sickness like typhoid. Beccaria and his nineteenth and twentieth century followers argued that punishment should be based on the findings of modern science. Criminologist Thorsten Sellin described the impact of Beccaria and his school of thought:

> [T]he struggle [about the death penalty] has been one between ancient and deeply rooted beliefs in retribution, atonement or vengeance on the one hand, and, on the other, beliefs in the personal value and dignity of the common man that were born of the democratic movement of the eighteenth century, as well as beliefs in the scientific approach to an understanding of the motive forces of human conduct, which are the result of the growth of the sciences of behavior during the nineteenth and twentieth centuries. If these newer trends of our thinking continue undisturbed the death

penalty will disappear in all the countries of Western culture sooner or later.[10]

Beccaria's position hit America's shores in 1787 when Dr. Benjamin Rush (a signer of the Declaration of Independence) published *An Enquiry into the effects of Public Punishment upon Criminals and upon Society*. The essay was the catalyst for the elimination of public punishments—flogging, whipping posts, public labor, etc.—and the creation of the first penitentiary. According to Rush and his followers, punishment should not be retributive but should be intended to reform the criminal. It was argued that a reformed convict would become a fruitful member of society. Prisoners would be penitent, and punishment would consist of "bodily pain, labor, watchfulness, solitude and silence" and "regular instructions in the principles and obligations of religion."[11]

Rush and his American intellectual heirs, like the European Enlightenment thinkers, assumed that all moral choices were determined by physical causes. Hence, science, not the death penalty, was the answer to ending vice. (One is reminded of the scientific claim of Pierre Cabanis, the father of physiological psychology, that all poetry and religion are products of the small intestines.)

The reader should be aware that the modern ideologues—Hobbes, Mill, Feurbach, Marx, Comte, Spencer, Dewey, Freud, Skinner—were influenced by the Cartesian view of a quantitative universe and its materialist conception of man; a view that was explained at length in chapter five of this book. For purposes of our discussion of the death penalty, it is important to recall that Descartes and his epigones deny that man has the freedom to choose between good and evil. Moral choice is nothing more than the tropism of an automation conditioned by various genetic, social, and historical contingencies. Crime is considered a disease, for which the criminal necessarily lacks responsibility, and criminology is the process of seeking cures. At its extreme, this view led to the thirty-two state laws that in the mid-1940s mandated the sterilization of criminals.

More recently rehabilitation has taken the form of educational, counseling, and psychotherapy programs provided by sociologists, social workers, and psychiatrists. Billions of federal, state, and local dollars were expended on such programs, even as Americans experienced ever increasing rates of crime. It was not until the 1990s, after a decade of beefed-up law enforcement (including the reinstatement of the death penalty in a number of states), that those rates began to decline.

Prominent psychiatrist Willard Gaylin, who has said that "nothing is wrong—only sick,"[12] has seen the writing on the wall. Even though he continues to believe that, from a psychiatric point of view, criminals are not responsible for their crimes, he still insists that the law should exact retribution:

> The new psychological definition of human beings that had each of us operating in a reality different from those around us destroyed the actual world. But is it the actual world that the law must operate in [in] order to preserve equity. In our search for individual justice we must not destroy the sense that we are living in a fair and just state. . . . A just society traditionally does some disservice to its individual members. The common good demands sacrifice of the individual. That is the lesson in the most moral of doctrines. The community under Jehovah is a community of law and justice, and yet the prophets may demand the ultimate sacrifices, even unto death, for the preservation of the law and the people of the law.[13]

# 22

# CAPITAL PUNISHMENT:
# THE CONSTITUTION
# AND THE SUPREME COURT

*[C]ontrary to abolitionist hopes and expectations, the Court did not invalidate the death penalty. It upheld it. It upheld it on retributive grounds. In doing so, it recognized, at least implicitly, that the American people are entitled as a people to demand that criminals be paid back, and that the worst of them be made to pay back with their lives. In doing this, it gave them the means by which they might strengthen the law that makes them a people, and not a mere aggregation of selfish individuals.*

Walter Berns

THE FOUNDING FATHERS, including Washington and Jefferson, accepted the death penalty as appropriate punishment for heinous crimes.[1] The Constitution itself recognizes capital punishment. The Fifth Amendment states

that "no person shall be held to answer for a capital or otherwise infamous crime, unless on a presentment or indictment of
a Grand Jury . . . nor shall any person be subject for the same
offense to be twice put in jeopardy of life or limb . . . nor be
deprived of life, liberty or property, without due process of law."
The double-jeopardy clause obviously assumes that there are
crimes for which, with due process of law, a man may lose his
life.

The Fourteenth Amendment, adopted in 1868, provides that
no state shall deprive any person of life, liberty or property without
due process of law, which simply reiterates the assumptions of
the Fifth. The Constitution assumed that capital punishment was
an option, but left it to the Congress and the states to decide
when it should be required or, presumably, if it should be abolished. The very first Congress, for instance, in "An Act for the
Punishment of Certain Crimes Against the United States," included punishment by death for numerous crimes.[2]

The usual arguments for banning the death penalty are and
have been based on the "cruel and unusual punishment" clause
of the Eighth Amendment. It is claimed that by its very nature
capital punishment is "cruel and unusual." This never occurred
to the nation's Founders for whom the phrase "cruel and unusual punishment" meant drawing and quartering, disembowel-
ing, burning at the stake—all forms of punishment still customary
in the eighteenth century—and it was these practices that they
were prohibiting.[3]

The definition of what constitutes a cruel and unusual punishment has depended upon the evolving standards of decency
that marked the progress of a maturing society.[4] As the perspective
of the populace changed over time, America's penal codes were
made less brutal and were restructured to limit the use of capital punishment in particular, and in general to ensure that all

punishments fit the crime. Degrees of murder were instituted, public executions were eliminated, the variety of capital statues was reduced, and juries were granted greater discretion in sentencing.

Throughout the nineteenth and early twentieth centuries, there was little constitutional debate about the validity of the death penalty itself, and those cases that did reach the Supreme Court were mostly concerned with procedural matters. It was not until the 1970s that the Court agreed to review cases in which plaintiffs argued that the death penalty is "cruel and unusual, " and therefore in violation of the Eighth and Fourteenth Amendments, and in *Furman v. Georgia* (1972) a divided Court (5-4) did indeed declare the use of death penalty to be unconstitutional. The Court ruled that extant federal and state death-penalty statutes were being administered in "an arbitrary or capricious manner," and summarily voided them all, although it left open the possibility that new statutes might be written that would pass constitutional muster.[5] It was a stunning break with legal tradition.

And if opponents of the death penalty were thrilled with the decision, they were ecstatic about a concurring opinion written by Justice William Brennan in which he held that any and all applications of punishment by death would be "cruel and unusual": "It is a denial of human dignity for the State arbitrarily to subject a person to an unusually severe punishment that society has indicated it does not regard as acceptable, and that cannot be shown to serve any penal purpose more effectively than a significantly less drastic punishment. Under these principles and this test, death is today a "cruel and unusual" punishment."[6] His statement left the door wide open for future challenges.

Reacting to the Court's ruling in *Furman*, thirty-five states enacted death-penalty statutes that they considered to be neither arbitrary nor capricious. In *Gregg v. Georgia* (1976) the Court upheld the state of Georgia's rewritten law, and executions resumed thereafter in a few states. In this and subsequent cases (including *Proffit v. Florida* and *Jurek v. Texas*) the Court attempted to apply the standard it established in Furman, and in doing so proved that the views of Justice Brennan were insupportable. The key to the decision in *Gregg* is the Court's determination to show that capital punishment is (or can be) a valid form of just retribution:

> It is apparent from the text of the Constitution itself that the existence of capital punishment was accepted by the Framers. . . . For nearly two centuries, this Court, repeatedly and often expressly, has recognized that capital punishment is not invalid per se. . . . And in *Trop v. Dulles*, 356 U.S., at 99, Mr. Chief Justice Warren, for himself and three other Justices, wrote: "Whatever the arguments may be against capital punishment, both on moral grounds and in terms of accomplishing the purposes of punishment . . . the death penalty has been employed throughout our history, and, in a day when it is still widely accepted, it cannot be said to violate the constitutional concept of cruelty."[7]

On the issue of retribution the court concluded:

> In part, capital punishment is an expression of society's *moral outrage* at particularly offensive conduct. This function may be unappealing to many, but it is essential in an ordered society that asks its citizens to rely on legal processes rather than self-help to vindicate their wrongs.
>
> The instinct for retribution is part of the nature of man, and channeling that instinct in the administration of criminal justice serves an important purpose in promoting the stability of a society governed by law. When people begin to believe that organized society is unwilling or unable to impose upon criminal offenders the punishment they "deserve," then there are sown the seeds of anarchy—of self-help, vigilante justice, and lynch law. . . .

Indeed, the decision that capital punishment may be the appropriate sanction in extreme cases is an expression of the community's belief that certain crimes are themselves so grievous an affront to humanity that the only adequate response may be the penalty of death.[8]

On the issue of deterrence the court stated that the "results simply have been inconclusive":

Statistical attempts to evaluate the worth of the death penalty as a deterrent to crimes by potential offenders have occasioned a great deal of debate. The results simply have been inconclusive. As one opponent of capital punishment has said:

"[A]fter all possible inquiry, including the proving of all possible methods of inquiry, we do not know, and for systematic and easily visible reasons cannot know, what the truth about this 'deterrent' effect may be."

. . . The inescapable flaw is . . . that social conditions in any state are not constant through time, and that social conditions are not the same in any two states. If an effect were observed (and the observed effects, one way or another, are not large) then one could not at all tell whether any of this effect is attributable to the presence or absence of capital punishment. A "scientific"—that is to say, a soundly based—conclusion is simply impossible, and no methodological path out of this tangle suggests itself. C. Black, *Capital Punishment: The Inevitability of Caprice and Mistake* 25-26 (1974).

Although some of the studies suggest that the death penalty may not function as a significantly greater deterrent than lesser penalties, there is no convincing empirical evidence either supporting or refuting this view. We may nevertheless assume safely that there are murderers, such as those who act in passion, for whom the threat of death has little or no deterrent effect. But for many others, the

death penalty undoubtedly is a significant deterrent. There
are carefully contemplated murders, such as murder for hire,
where the possible penalty of death may well enter into the
cold calculus that precedes the decision to act. And there
are some categories of murder, such as murder by a life
prisoner, where other sanctions may not be adequate.[9]

Walter Berns, a noted authority on capital punishment, had this
reaction to the *Gregg* decision:

> [C]ontrary to abolitionist hopes and expectations, the Court
> did not invalidate the death penalty. It upheld it. It upheld
> it on retributive grounds. In doing so, it recognized, at least
> implicitly, that the American people are entitled *as a people*
> to demand that criminals be paid back, and that the worst
> of them be made to pay back with their lives. In doing this,
> it gave them the means by which they might strengthen the
> law that makes them a people, and not a mere aggregation
> of selfish individuals.[10]

Since *Gregg* the Supreme Court has ruled twenty-two times
on procedural and substantive rules concerning the death penalty's
application. More than 450 criminals have been executed and
over 3,500 currently reside on "death row."

# 23

## SOCIETY'S MORAL OUTRAGE

*Capital Punishment is an expression of society's moral outrage at particularly offensive conduct.*

Potter Stewart
Associate Justice of the Supreme Court

O PPONENTS OF THE DEATH PENALTY frequently (and all too successfully) entice proponents to discuss the issue on the abolitionist's own terms. During New York's 1994 election campaign, many opponents of the death penalty, aided and abetted by a biased media, attempted to limit the death-penalty debate to the question of deterrence. Frankly, that discussion is specious: the evidence is inconclusive, contradictory, and fundamentally irrelevant.

Punishment is inflicted by the state and deserved by the criminal *because of the crime*; it is not primarily intended "to de-

ter others from similar acts."[1] If deterrence were the sole purpose of punishment, it is possible that criminals might actually
commit more crimes, since, in the words of Rousseau, "they will
soon enough discover the secret of how to evade the laws."[2]

Hobbes, Beccaria, and their intellectual heirs believe only
in positive law. They deny the existence of moral (or natural)
law. Since man is moved only by the instinct for self-preservation, his acts, including criminal acts, are neither moral nor
immoral, right nor wrong. What we call crimes, the positivists
call violations of people's self-interests, and they promise that
when men submit to self-interest crime will disappear. This is
the great fallacy of enlightened social engineers. In his *Notes from
the Underground*, Fyodor Dostoevsky exposed their illusions:

> But these are all golden dreams. Oh, tell me, who was it
> first announced, who was it first proclaimed, that man only
> does nasty things because he does not know his own in-
> terests; and that if he were enlightened, if his eyes were
> open to his real normal interests, man would at once cease
> to do nasty things, would at once become good and noble
> because, being enlightened and understanding his real ad-
> vantage, he would see his own advantage in the good and
> nothing else, and we all know that not one man can, con-
> sciously, act against his own interests, consequently, so to
> say, through necessity, he would begin doing good? Oh, the
> babe? Oh, the pure innocent child![3]

Despite the predictions of the social engineers, human
behavior has not fundamentally matured or progressed since
Dostoevsky wrote those words at the time of the American Civil
War. Crime has been rampant throughout America, and we are
plagued by the likes of Richard Speck, Charles Manson, Gary
Gilmore, Sirhan Sirhan, Jeffrey Dahmer, and Timothy McVeigh.

Men must be held responsible for their actions. In communities based on natural law and supplemented by positive laws,

men freely obey laws and are trusted to do so. A criminal who violates that trust violates the common good. All the members of the community are victims because their dignity is violated. And this community of victims who are morally outraged demand justice because if they did not, one could conclude that the individual members cared only about themselves and not the community.

The primary purpose of the state is to maintain order and to perform the duty required in the use of power to prevent and punish criminal activity. The state, having received its power from God through the natural law, has the right to create the means for attaining justice. The death penalty is a means of retributive justice to re-establish the balance of *outraged* justice against the most serious of crimes. As Justice Stewart declared in *Gregg:* "Capital Punishment is an expression of society's *moral outrage* at particularly offensive conduct."[4]

Outrage is a legitimate form of moral indignation. When American forces entered the liberated Nazi death camps, they were outraged. General Eisenhower vented his outrage by ordering the German citizens living in the vicinity to view the effects of their ideological madness, and to bury the dead. The Nuremberg trials were held to express the world's moral outrage and the convicted Nazi thugs received just retribution—the hangman's noose.

Retributive justice cannot be abolished because justified punishments are *inherently* retributive. A person found guilty of a crime is punished to re-establish the balance of justice.[5] Retributive punishment can only be administered by the state and not by individual citizens for one reason, because only the state is neutral. It does not act out of vengeance as individuals easily may. In fact, retribution is the *primary* reason for punishment. That is why only the state may administer justice. Revenge and vengeance compound an evil, but justice is a good act. To throw out retribution as a so-called relic of the Dark Ages is wrong. Writing in *Crime and Human Nature*, James Q. Wilson agrees:

> Defenses of retribution center on the notion of equity. An offender has violated an implicit social contract that ties the members of a community together. The contract gives the members rights to goods obtained only in legal ways. Goods gained illegally violate the equity principle, because an illegal act is by definition not an acceptable input. Breaking the law gives an offender an unfair advantage in the pursuit of reward. Hence the criminal's ratio of gains to inputs will seem to other parties to the social contract to be unacceptably high. The criminal is therefore said to "owe a debt" to society. Punishment as retribution balances the books. Crimes of violence also earn "goods" in a metaphoric sense—they are, at least at the moment they occur, rewards exceeding what has been earned by legitimate inputs—and are similarly reckoned as a debt requiring repayment.[6]

The Committee for the Study of Incarceration, chaired by former U.S. Senator Charles Goodell, concluded that the "principle of commensurate deserts in our opinion is a requirement of justice. . . The social benefits of punishing do not alone justify depriving the convicted offender of his rights, it is also necessary that the deprivation be deserved."[7]

Noted Kantian philosopher Jeffrie Murphy wrote in *Retribution, Justice and Therapy* (1979) that "a retributive theory of justice . . . is the only morally acceptable theory of punishment. . . . [T]he twentieth century's faddish movement toward a 'scientific or therapeutic' response to crime runs grave risks of undermining the foundation of justice."[8]

Although some may quip that the death penalty is a deterrent in the sense that the executed criminal is certainly deterred from committing additional crimes, the fact is capital punishment is, by its very nature, retributive.

It has been demonstrated in this section that to maintain the common good of the community, justice requires a duly authorized government to inflict punishment that fits the crime. To restore the balance of justice, the civil government has the power of the sword to punish incorrigibly dangerous malefactors.

When a person commits an act of horrendous malice, the state is permitted to cut him off from society. Thomas Aquinas reminded us that just as it is reasonable to cut off a diseased member of the human body when this member threatens the welfare of the whole body, so it is reasonable to permit the body politic to cut off a bad member of society for the sake of the good of the whole.[9]

While taking of innocent life is always morally wrong, it cannot be argued that the taking of a life is always wrong—otherwise it would be wrong to defend oneself against an unjust aggressor. The natural law gives a person the right to kill in self-defense just as it gives the state the power to make war against—and kill—an unjust invader who threatens the common good.

In the case of the Oklahoma City bombing, Mr. McVeigh violated the common good. He challenged the authority of the state and destroyed innocent life. When proven guilty, the jury of his peers, rightfully inflicted the sentence of death as a means to re-establish the balance of outraged justice and for the preservation and well-being of the state.

The Nuremberg Trials revealed to the world the injustices, the cruelties, the viciousness, the human degradation perpetrated by the Nazi regime. As Robert Jackson, the American prosecutor, stated, "The wrongs which we seek to condemn and punish have been so calculated, so malignant, and so devastating, that civilization cannot tolerate their being ignored, because it cannot survive their being repeated. That four great nations, flushed with victory and stung with injury, *stay the hand of vengeance* and voluntarily submit their captive enemies to the judgment of the law is one of the most significant tributes that Power has ever paid to Reason."[10]

The trials proved "beyond all peradventure" a conspiracy of evil, and most of the defendants were condemned to the gallows. Their execution did not violate their human dignity, but it was a means of restoring lost dignity to the millions of victims of Nazi atrocities.

The death penalty, at least for premeditated murder, does not operate in opposition to human dignity. Rather, it was established in respect of human dignity, which is derived from the *imago Dei*—the image of God—within each human being. Capital punishment is the ultimate compliment to the human dignity of both victim and murderer; it implies the most pro-human stance possible. There are, of course, forms of capital punishment which violate human dignity and worth, but the basic concept is entirely compatible with God's standards of human worth. On the other hand, a penalty of ten years in prison for premeditated murder devalues human life by saying that the victim's life was worth only ten years of penalty. Equity and justice demand a punishment that matches the crime.[11]

The death penalty remains an essential part of the state's criminal code and helps to ensure the preservation of the common good and the dignity of its citizenry.

# NOTES

## 1. The American Credo

1. Quoted from Thomas Jefferson's original draft as found in Sheldon, *The Political Philosophy of Thomas Jefferson* (Baltimore, 1993), p. 43.

2. Quoted in Sheldon, p. 34.

3. Sheldon, p. 30.

## 2. Natural Law on Trial: Nuremberg and the Higher Truth

1. Conot, *Justice at Nuremberg* (New York, 1983), p. 10.

2. Ibid., p. 11.

3. Ibid., p. 23.

4. Ibid., p. 11.

5. Sereny, *Albert Speer: His Battle with Truth*, (New York, 1995), p. 564.

6. Conot, p. 79.

7. Rice, *Beyond Abortion: The Theory and Practice of the Secular State* (Chicago, 1979), p. 11.

8. Azar, *Twentieth Century in Crisis: Foundations of Totalitarianism*, (Iowa, 1990), p. 200.

9. Friedlander, *Nazi Germany and the Jews: Volume 1*, (New York, 1997), p. 142.

10. Ibid., p. 152.

11. Ibid., p. 120.

12. Ibid., p. 27.

13. Ibid., p. 33.

14. Ibid., p. 243.

15. Kreeft, *Making Choices* (Ann Arbor, 1990) p. 38.

16. Azar, p. 204.

17. Conat, p. 344.

18. Sereny, Page 577.

19. Azar, Page 203.

20. Tusa, *The Nuremberg Trial* (New York, 1983) p. 155.

## 3. Discovering Natural Law: An Historical Overview

1. Rice, *50 Questions on the Natural Law* (San Francisco, 1993), pp. 29-30.

2. Rice, *Beyond Abortion* (Chicago, 1979), p. 32.

3. Aquinas, *Summa Theologica*, I-II q. 92, art. 2.

4. Rice, *50 Questions on the Natural Law*, p. 30.

5. Rommen, *The Natural Law* (St. Louis, 1947), p. 6.

6. Rice, *50 Questions on the Natural Law*, p. 30.

7. Sigmund, p. 14.

8. Ibid., p. 18.

9. Sigmund, p. 22.

10. Rommen, p. 23.

11. Ibid., p. 35.

12. Strauss, *History of Political Philosophy* (Chicago, 1963), p. 220.

13. Sigmund, pp. 52-54.

14. Ibid.

15. Rice, *50 Questions on the Natural Law*, p. 30.

16. Sigmund, p. 99.

17. Sigmund, p. 101.

18. Rice, *50 Questions on the Natural Law*, p. 33.

19. Azar, Page 264.
20. Lincoln, *Speeches and Writings: 1859-1865* (New York, 1984).
21. Azar, p. 264.
22. Wilson, *The Moral Sense* (New York, 1993), p. 12.

## 4. The Politics of Natural Law: The Person and the Common Good

1. Baker, *Homiletic and Pastoral Review*, Volume LXXXVI, 5, p. 80.
2. Adler, *The Common Sense of Politics* (New York, 1996), pp. 30-31.
3. Azar, p. 214.
4. Ibid.
5. Ibid.
6. Ibid., p. 234.
7. Novak, *Free Persons and the Common Good* (New York, 1989), p. 124.
8. Fitzgerald, *The Catholic Lawyer*, April, 1956.
9. Ibid.
10. *The Catholic Encyclopedia*, p. 762.
11. Ibid.
12. Azar, p. 237.
13. Sheen, *Liberty, Equality, and Fraternity* (New York, 1938) p. 139.
14. Sheen, p. 144.
15. Maritain, *The Person and the Common Good* (Indiana, 1966), pp. 52-53.
16. Fitzgerald, *Catholic Lawyer* (April, 1956).

## 5. Natural Law Under Fire: Ideological Opponents Old and New

1. Azar, pp. 27-28.
2. Ibid.
3. Hobbes, *Leviathan* (London, 1904), Chapter 13.
4. Ibid.
5. Holland, *Jurisprudence* (Oxford, 1937), p. 54.

6. Rustow, *Freedom and Domination* (Princeton, 1980), p. 523.

7. Fogothey, *Ethics in Theory and Practice* (St. Louis, 1959), p. 116.

8. Rommen, pp. 81-82.

9. Ibid., p. 112.

10. Ibid.

11. Sabine, *A History of Political Theory* (New York 1963), pp. 604-605.

12. Fogothey, p. 63.

13. Wexler, "The Moral Confusions in Positivism, Utilitarianism, and Liberalism," *The American Journal of Jurisprudence*, Volume 38 (1993), p. 121.

14. Ibid.

15. Grisez and Shaw, *Beyond The New Morality* (Indiana, 1988), p. 111.

16. Wexler, p. 133.

17. Rommen, p. 113.

18. Ibid., pp. 128-129.


6. THE NATURAL LAW AND THE UNITED STATES OF AMERICA

1. Sheldon, pp. 56-57

2. Fogothey, p. 386.

3. Azar, p. 290.

4. Ibid., p. 57.

5. Ibid., p. 59

6. Ibid., p. 111.

7. Ibid., p. 9.

8. Ibid., p. iv.

9. Rice, *Beyond Abortion*, p. 41.

10. Rice, *50 Questions on the Natural Law*, p. 34.

11. Azar, p. 216.

12. Ibid., p. 212.

13. Ibid., p. 213.

14. Ibid.

15. Sigmund, p. 104.

16. Rice, *50 Questions on the Natural Law*, p. 34.

17. Azar, p. 264.

18. Ibid.

### 7. THE NATURAL LAW: AMERICAN IDEOLOGICAL OPPOSITION

1. James, *Letters* (New York, 1920), Volume 2, p. 279.

2. James, *Pragmatism* (New York, 1931), p. 51.

3. Ibid., p. 53.

4. Ibid., p. 63.

5. Ibid., p. 64.

6. Ibid., pp. 222-223.

7. Sumner, *Folkways* (New York, 1913), p. 2.

8. McKinnon, "Natural Law and the Gentiles," *Catholic University Law Review* (1955), p. 1.

9. Wexler, p. 122.

10. Frankfurter, *Harvard Law Review*, (1931), p. 717.

11. Holmes, *Holmes-Pollock Letters* (New York, 1941), p. 163.

12. Holmes, *Holmes Book Notices, Uncollected Letters and Papers* (New York, 1936), p. 165.

13. *Holmes-Pollock Letters*, p. 122.

14. Ibid., p. 252.

15. Holmes, "Ideals and Doubts," *Illinois Law Review* (1915), p. 2.

16. Holmes, *Collected Legal Papers* (New York, 1920), p. 310.

17. Holmes, *Book Notices*, pp. 187-188.

18. Ibid., p. 181.

19. Lerner, *The Mind and Faith of Justice Holmes*, (New York, 1943), p. 50.

20. Ibid., p. 73.

21. Holmes, "Natural Law" (*Harvard Law Review*, August 1918), p. 40.

22. Holland, p. 83.

23Holmes, "Ideals and Doubts," p. 3.

24. *Holmes-Pollock Letters*, p. 200.

25. Ibid., p. 178.

26. Ibid., p. 163.

27. Ibid., p. 36.

28. Holmes, "Natural Law."

29. Holmes, *Book Notices*, pp. 187-188.

30. Lerner, p. 50.

31. Ibid.

32. *Holmes-Pollock Letters*, p. 212.

33. Ibid., p. 307.

34. Ibid., p. 171.

35. Lief, editor, *The Dissenting Opinions of Mr. Holmes* (New York, 1929), p. 50.

36. Holmes, *Collected Legal Papers*, p. 169.

37. Cohen, *Columbia Law Review* (1935), p. 843.

38. Bingham, *My Philosophy of Law* (New York, 1941), p. 16.

39. Weintraub, *Natural Law and Justice*, (Cambridge, 1987), p. 118.

40. MacIntyre, *After Virtue* (South Bend, 1984), p. 246.

41. Budzizewski, *True Tolerance*, (New Jersey, 1992), Page 4.

42. *New York Times* (July 15, 1991).

## 8. The Need for Natural Law

1. Holmes, *Collected Legal Papers*, p. 200.

2. Miner, *The Concise Conservative Encyclopedia* (New York, 1996) p. 194.

3. Rice, *Beyond Abortion*, p. 49.

4. *Dennis v. United States*, 71 Sup. Ct. 857, 866 (1951).

5. Morley, "Affirmation of Materialism," *Barron's* (June 18, 1951), p. 3.

6. Sabine, p. 902.

7. Ibid.

8. Azar, p. 180.

9. Ibid., p. 170.

10. Ibid., p. 181.

11. Ibid., p. 196.

12. Ibid., p. 172.

13. Johnson, *Modern Times* (New York, 1991), pp. 289-290.

14. Azar, p. 170.

15. Lucey, "Jurisprudence and the Future Social Order," *Social Science* (July, 1941), p. 1211.

16. Lucey, "Holmes, Liberal, Humanitarian, Believer In Democracy," *Georgetown Law Journal* (May, 1951), p. 548.

17. Palmer, "Hobbes, Hitler and Holmes" (*A.B.A. Journal*, November, 1945), p. 573.

18. Rice, *50 Questions on the Natural Law*, p. 79.

19. Lucey, *Georgetown Law Journal* (1951), p. 553.

20. Lucey, *Georgetown Law Journal* (1942), p. 523.

21. Rice, *Beyond Abortion*, p. 51.

22. Ibid., p. 47.

23. Neuhaus, *The Human Life Review* (Winter, 1987), p. 81.

24. *New York Times* (July 15, 1991).

25. Cromartie, *A Preserving Grace: Protestants, Catholics and Natural Law* (Washington, DC, 1992), p. 29.

26. Gregg, "The Pragmatism of Mr. Justice Holmes," *Georgetown Law Journal* (1943), p. 288.

27. Ibid., p. 292.

28. Lucey, *Georgetown Law Journal* (1951), p. 562.

## 9. Euthanasia through the Centuries

1. *The New England Journal of Medicine* (November 28, 1996).

2. Marx, *And Now Euthanasia* (Washington, DC, 1985), p. 10.

3. Ibid. p. 11.

4. Augustine, *The City of God* (New York, 1972) Book 1, Chapter 17.

5. Barry, "The Development of the Roman Catholic Teachings on Suicide," *Notre Dame Journal of Law, Ethics and Public Policy*, Volume 9 (1995), p. 475.

6. Ibid.

7. Aquinas, *Summa Theologica*, II-II, Question 64.

8. Marx, p. 12.

9. Ibid.

10. Ibid., p. 13.

11. Vatican Congregation for the Doctrine of the Faith, *Declaration on Euthanasia*, (1981), p. 8.

12. Ibid.

13. O'Donnell, *Medicine and Christian Morality* (New York, 1978), p. 45.

14. Marx, p. 16.

15. Hofstadter, *Social Darwinism in American Thought* (New York, 1955), p. 103.

16. Ibid. p. 51.

17. *Encyclopedia of Philosophy*, Volume 2, p. 305.

18. Kimbrell, *The Human Body Shop* (San Francisco, 1993), p. 253.

19. Fulk, "Euthanasia: The Gentle Death," *The Catholic Lawyer*, Volume 35, No. 2, p. 151.

20. Ibid.

21. Proctor, *Racial Hygiene* (Cambridge, 1988), p. 178.

22. Ibid., p. 180.

23. Ibid.

24. Kuhl, *The Nazi Connection: Eugenics, American Racism* (New York, 1994), p. 86.

25. Proctor, p. 180.


10. The German Euthanasia Experience: The Final Solution

1. Lipton, *The Nazi Doctors* (New York, 1986), p. 46.

2. Ibid.

3. Proctor, p. 25.

4. Ibid., pp. 178-179.

5. Lipton, p. 46.

6. Friedlander, *The Origins of Nazi Genocide*, (Chapel Hill, 1995), p. 14.

7. Ibid., p. 15.

8. Proctor, p. 28.

9. Ibid., p. 64.

10. Ibid., p. 70.

11. Ibid.

12. Ibid.

13. Conat, *Justice at Nuremberg* (New York, 1983), p. 205.

14. Fischer, *Nazi Germany: A New History* (New York, 1995), p. 389.

15. Sereny, *Albert Speer: His Battle with Truth* (New York, 1995), p. 196.

16. Gallagher, *By Trust Betrayed* (Virginia, 1995), p. 260.

17. Proctor, p. 183.

18. Lipton, p. 142.

19. Friedlander, p. 115.

20. Breitman, *The Architect of Genocide: Himmler and the Final So-lution* (New York, 1991), p. 90, and Smith, *Forced Exit*, p. 80.

21. Friedlander, p. 20.

22. Kinter, *The Haldamar Trial* (London, 1949), pp. 22 and 234.

23. "Extracts from the Closing Brief for Defendant Karl Brandt (The Medical Case), Volume 1, p. 834

24. Rice, *Beyond Abortion* (Chicago, 1979), pp. 131-132

25. Muggeridge, *Vintage Muggeridge* (Michigan, 1985), p. 63

26. *The American Journal of Nursing*, 1973.

27. Marx, p. 21

## 11. Anything Goes: The Dutch Euthanasia Experience

1. Huizinga, *Dutch Civilization in the Seventeen Century and Other Essays*, p. 121.

2. Hendrin, *Seduced by Death*, p. 75.

3. "Physician-Assisted Suicide and Euthanasia in the Netherlands," *Report of the Subcommittee on the Constitution, Judiciary Committee of the House of Representatives*, (September, 1966), p. 4.

4. Ibid., p. 5.

5. Ibid., p. 6.

6. Gomez, *Regulating Death* (New York, 1991), p. 32.

7. Report of Subcommittee on the Constitution, p. 1.

8. Ibid., p. 12.

9. Hendrin, *Seduced by Death*, pp. 68-69.

10. Ibid., p. 48.

11. Ibid., p. 49.

12. Ibid., p. 76.

13. Ibid., p. 73.

14. Ibid., p. 80.

15. Report of the Subcommittee of the Constitution, p. 17.

16. Hendrin, *Seduced by Death*, Page 94.

17. Fenigsen, "Physician-Assisted Suicide and Euthanasia in the Netherlands," *Issues in Law and Medicine*, Volume 11, (1995).

18. Hendrin, *Seduced by Death*, p. 240.

19. Ibid., p. 110-111.

20. Ibid., p. 98.
21. Ibid., p. 105.

## 12. THE SLIPPERY SLOPE: THE AMERICAN EUTHANASIA EXPERIENCE

1. Alvarez, "Death with Dignity," *The Humanist* (September 1971), p. 12.

2. *The Florida Sun* (January 11, 1973).

3. "Memorandum on Additional Cost-Savings Initiatives," Department of Health Education and Welfare (June 4, 1977), pp. 8-9.

4. Brennan, William, *The Abortion Holocaust* (St. Louis, 1983), p. 83.

5. "Infant Euthanasia: Crime or Compassion?" *Milwaukee Journal*, (June 15, 1981).

6. "It's Over Debbie," *Journal of the American Medical Association*, January (1988).

7. *U.S. News and World Report* (February 22, 1988).

8. *New York Times*, February 24, 1988.

9. *The New England Journal of Medicine* (March 1991).

10. American Medical Association Council on Ethical and Judicial Affairs, Opinion 2.5 (1986).

11. *New York Times* (May 23, 1996).

12. *New York Times* (July 15, 1996).

13. Smith, *Forced Exit* (New York: 1992), pp. 116-117.

14. *Newsday* (March 24, 1996).

15. Kevorkian, *Prescription Medicide: The Goodness of a Planned Death* (New York, 1991), pp. 202-203.

16. *Boston Globe* (August 17, 1996).
    *Washington Post* (August 21, 1996).
    *New York Times* (August 17, 1996 and August 26, 1996)

17. *Newsday* (August 20, 1996).

18. *New York Times* (August 29, 1996).

19. *The Washington Post* (June 26, 1996).

20. *USA Today* (July 30, 1996).

13. Courting Death: Recent Legal Decisions
on Assisted Suicide

1. Smith, pp. 45-50.

2. Senander, *The Living Will* (St. Paul, 1993), p. 22.

3. Smith, p. 50.

4. American Medical Association Council on Ethics and Judicial Affairs, Opinion 2.20 (1994).

5. Schaffner, *Critical Care Medicine* (October, 1988).

6. Singer, *Rethinking Life and Death* (New York, 1995), p. 80.

7. *Compassion in Dying, Inc. v. State of Washington*, F. 3rd 49 (1995).

8. *Compassion in Dying, Inc. v. State of Washington*, 79F, 3rd 790, 9th Circuit (1996).

9. Ibid.

10. Ibid.

11. Ibid.

12. Ibid.

13. Ibid.

14. Ibid.

15. Ibid.

16. Ibid.

17. Ibid.

18. Ibid.

19. *Vacco v. Quill*, Petition For a Writ of Certiorari (May 15, 1996), No. 95-1858.

20. Ibid.

21. *Vacco v. Quill*, Petition For a Writ of Certiorari (May 15, 1996), Appendix 1A.

22. Ibid.

23. Ibid., Appendix 63A

24. Ibid.

25. Ibid.

26. See April 3, 1996 editions of the *New York Times, Washington Post, Newsday, New York Daily News,* and *New York Post.*

27. Ibid.

28. Ibid.

29. Ibid.

30. *Los Angeles Times* (April 3, 1996).

31. See April 8, 1996 editions of *New York Times, Newsday,* and *New York Post.*

32. *Chicago Tribune* (November 10, 1996).

33. *Vacco v. Quill*, Brief for Respondents, On Writ of Certiorari to the U.S. Court of Appeals for the Second Circuit (December 10, 1996).

34. Ibid.

35. Ibid.

36. Ibid.

37. Ibid.

38. Ibid.

39. Ibid.

40. Ibid.

41. *Vacco v. Quill*, Official Transcript Proceedings Before The Supreme Court of the United States (January 8, 1997).

42. *Vacco v. Quill*, Reply Brief for Petitioners. Vacco and Pataki On Writ of Certiorari to the U.S. Court of Appeals for the Second Circuit (December 24, 1996).

43. Ibid.

44. Ibid.

45. *New York Times* (June 27, 1997)

46. Ibid.

47. *Vacco v. Quill*, U.S. Supreme Court (June 26, 1997).

48. Ibid.

49. Ibid.

50. Ibid.

51. *New York Times* (June 27, 1997).

52. Ibid.

53. Ibid.

54. Ibid.

55. Ibid.

56. Ibid.

57. Ibid.

## 14. THE DEBATE GOES ON:
## DOES A PERSON HAVE THE "RIGHT TO DIE"?

1. Mill, *Utilitarianism* (New York, 1931), pp. 9-10.

2. Bentham, *Principles of Morals and Legislation* (Oxford, 1907), p. 102.

3. Barry, p. 466.

4. *Brophy v. New England Sinai Hospital,* 497 N.E. 2nd 626 (1986).

5. *Bouvia v. Superior Court,* 225 Cal. Rptr. 297, Cal. Ct. App. (1986).

6. Ibid.

7. Ibid.

8. *Cruzan v. Director, Missouri Department of Health,* 497 U.S. 261 (1990).

9. *Brophy v. New England Sinai Hospital*

10. Ibid.

11. Ibid.

12. New York Task Force on Life and The Law, *When Death is Sought: Assisted Suicide and Euthanasia in the Medical Context* (1994). [The quotes and summary of findings are from the Executive Summary of the Task Force's Report.]

13. Ibid.

14. Gula, *Euthanasia,* (New Jersey 1994), p. 35.

15. Ibid.

16. Hentoff, "Death as a Way to Cut Health Care Costs," *Village Voice* (April 19, 1994).

17. "McNeill-Lehrer Newhour," on WNET-New York (March 18, 1994).

18. Hendrin, "Dying of Resentment," *New York Times* (March 21, 1996).

19. Weaver, *The Ethics of Rhetoric* (Chicago, 1965)

20. Hendrin, *Seduced by Death,* p. 224

21. Senander, p. 12

22. Smith, p. 229.

## 15. THE SCIENCE OF GOOD BIRTH

1. Kevles, *In the Name of Eugenics* (New York, 1985), p. 3.
2. Ibid, p. 4.
3. Azar, *Philosophy and Ideology* (Iowa, 1984), p. 18.
4. Ibid., p. 122.
5. Ibid., p. 46.
6. Ibid.
7. Ibid., p. 47.
8. Haller, *Eugenics* (New Jersey, 1984), Page 4.
9. Azar, p. 53.
10. Ibid.
11. Smith, *The Eugenics Assault On America* (Virginia, 1993), p. 1.
12. Ibid., p. 2.
13. Haller, p. 3.
14. Bier, *Human Life* (New York, 1985), p. 107.
15. Kevles, p. 33.
16. Ibid., p. 3.
17. Haller, p. 19.
18. Kevles, p. 92.
19. *The Holmes-Laski Letters Vol. II*, (Cambridge, 1953), p. 939.
20. Webb, *The Decline of the Birth Rate* (London, 1925), pp. 39-40.
21. Haller, p. 51.
22. Parmet, *Nixon and His America*, (New York, 1991), p. 4.
23. Marlin, "Is the Republican Party Losing Its Social Constituency and Thus The Nation?" *Crisis* (May 1990).
24. Ibid.
25. Ibid.
26. Dyer, *Theodore Roosevelt and the Idea of Race* (Louisiana, 1980), p. 158.
27. Ibid., p. 160-161.
28. Marlin.
29. Kevles, p. 21.
30. Ibid., pp. 45-47.
31. Chase, *The Legacy of Malthus* (Chicago, 1980), p. 111.
32. Kevles, p. 54.

33. Chase, p. 64.

34. Smith, p. 17.

35. Hasian, *The Rhetoric of Eugenics in Anglo-American Thought* (Georgia, 1996), p. 38.

36. Smith, p. 5.

37. Lipton, *The Nazi Doctors*, p. 23.

38. Chase, p. 16.

39. Ibid., pp. 94-96.

40. Osborn, *Preface to Eugenics* (New York, Harper & Row), p. 35.

41. Ibid., p. 298.

## 16. THE GERMAN EUGENICS EXPERIENCE

1. Lipton, p. 129.

2. Ibid., p. 24.

3. Proctor, p. 103.

4. Ibid., pp. 269-302.

5. Lukas, *The Forgotten Holocaust* (Kentucky), pp. 37-38.

6. Ibid., p. 39.

7. Stocking, *Race, Culture and Evolution* (New York, The Free Press), pp. 60-62.

8. Chase, pp. 94-97 and p. 163.

9. Ibid., p. 163.

10. Ibid., pp. 94-96.

11. Proctor, p. 98.

12. Chase, p. 635, note 11.

13. Proctor, p. 103.

14. Chase, p. 351.

15. Ibid., p. 349.

16. Ibid., p. 301.

17. Ibid., p. 634, note 9.

18. Proctor, p. 15.

19. Ibid., p. 17.

20. Ibid., p. 38.

21. Ibid., p. 22.

22. Ibid., p. 28.

23. Ibid., p. 45.

24. Ibid., p. 45.

25. Ibid.

26. Azar, p. 118.

27. Kuhl, *The Nazi Connection* (Oxford, 1994), p. 101.

28. Smith, p. 9.

29. Kuhl, p. 100.

17. THE NEW EUGENICS: GENETIC MANIPULATION AND CLONING

1. Ibid., p. 105.

2. Chase, p. 362 ff.

3. Ibid., p. 375.

4. Ibid., p. 388.

5. Ibid., p. 407.

6. Ibid.

7. Ibid., p. 482.

8. Neuhaus, *The Human Life Review* (Winter, 1987), p. 81.

9. Smith, *Daedalus*, (Summer 1989), p. 91.

10. Rini, *Beyond Abortion: A Chronicle in Fetal Experimentation* (New Jersey) pp. 33-34.

11. *Insight* (July 11, 1988), p. 14.

12. Neuhaus, p. 98.

13. Ibid., p. 96.

14. The *New York Times* (January 13, 1988).

15. *Daedalus*.

16. Kimbrell, *The Human Body Shop: The Engineering and Marketing of Life.* (New York, 1995), p. iii.

17. Ibid., p. 1.

18. Ibid., p. 13.

19. Ibid., p. 14.

20. Krauthammer, *Washington Post* (April 3, 1992).

21. Kimbrell, p. 20.

22. Ibid., pp. 30-31.

23. McNulty, *Chicago Tribune* (July 27, 1987).

24. Kimbrell, pp. 34-35.

25. Ibid., p. 33.

26. Ibid., pp. 35-36.

27. Ibid., p. 37.

28. Krauthammer.

29. Ibid.

30. Kimbrell, p. 44.

31. Ibid., p. 49.

32. Ibid., p. 70.

33. Ibid., p. 78.

34. Ibid., p. 73-74.

35. *The Broadside,* George Mason University.

36. Kimbrell, p. 88.

37. Ibid., p. 90.

38. Edmiston, *Glamour* (November 1991), p. 237.

39. *Anna Johnson v. Mark and Crispina Calvert,* Superior Court of the State of California for the County of Orange, No. x 63, 31, 90.

40. Harrison, *Chicago Tribune* (October 8, 1990).

41. Kimbrell, p. 120.

42. Virshup, *New York Magazine* (July 27, 1987).

43. Ibid.

44. Kimbrell, p. 125.

45. Kobilak and Gorner, *Chicago Tribune* (March 3, 1991).

46. Kimbrell, p. 129.

47. Tell, *The Weekly Standard* (September 15, 1997).

48. Cable News Network (CNN), Internet news story (February 24, 1997).

49. *Time* (March 10, 1997).

50. Fletcher, *The Ethics of Genetic Control* (New York, 1974), p. 71.

51. Kimbrell, p. 215.

52. Ibid., p. 217.

53. Hodgkinson, *The Sunday Times of London* (March, 1992).

54. Kimbrell, p. 214.

55. *New York Post* (March 4, 1997).

56. *Time* (March 10, 1997).

57. *New York Times* (March 5, 1997)

58. Ibid., (May 18, 1997)

59. Ibid., (June 8, 1997)

60. Kimbrell.

61. Ibid.
62. United States Catholic Conference, news release, June 13, 1997.

## 18. Does Human Cloning Violate the Dignity of the Person?

1. Azar, *Man: Computer, Ape or Angel* (Massachusetts, 1989), p.23.
2. Ibid.
3. Ibid., p. 24.
4. Ibid., p. 26.
5. Ibid., Page 29.
6. Fletcher, p. 156.
7. Kimbrell, p. 243.
8. Fletcher, p. 154.
9. Bonnici, "Cloning Technology and the Traditional Family," U.S. Capital Room SC-5 (June 24, 1997).
10. Ibid.
11. Proctor, p. 227.
12. Lipton, pp. 279 and 469.
13. Krutch, *The Measure of Man* (New York, 1997), p. 88.
14. Ibid.

## 19. Merely Vengeance or Morally Justifiable?

1. Bedau, *The Death Penalty in America* (New York, 1997) pp. 36-38.
2. Coyne and Entzepoth, *Capital Punishment and the Judicial Process* (North Carolina, 1995) pp. 79-83.
3. Polling statistics reported in *U.S. News and World Report* (June 16, 1997), p. 27.
4. *U.S. News and World Report,* "The Place for Vengeance" (June 16, 1997), pp. 25-32.
5. Ibid.
6. Ibid.
7. *New York Times* (June 4, 1997).
8. *The Boston Globe* (June 14, 1997).

9. *The Norwich Bulletin* (June 14, 1997).

20. THE ORIGINS OF THE DEATH PENALTY

1. Aquinas, *Summa Theologica*, II-II, 64.
2. *The Code of Maimonides* (Connecticut, 1949), Book 14, pp. 150-151.
3. Rice, *50 Questions on the Natural Law*, (San Francisco, 1993), p. 58.
4. Taken from the text of the Cardinal's sermon delivered at the New York Police Department Holy Name Mass as reported in the *New York Daily News* (March 31, 1996).
5. *Catechism of the Catholic Church* (San Francisco, 1994), p. 546.
6. *Code of Canon Law* (Canon Law Society of America, 1983) p.119.

21. OPPOSITION TO THE DEATH PENALTY: IDEOLOGICAL ORIGINS

1. Beccaria, *On Crimes and Punishment* (Indianapolis, 1963), p. 5.
2. Hobbes, Philosophical Rudiments Concerning Government and Society, Volume 2 (London, 1841), p. 196.
3. Quoted in Fagothey, *Ethics in Theory and Practice* (St. Louis, 1959) p. 384.
4. Hobbes, *Leviathan,* Chapter XIII (London, 1841), pp. 157-158
5. Ibid., p. 154.
6. Beccaria, p. 45.
7. Strauss, *Natural Right and History* (Chicago, 1953), pp. 190-191.
8. Quoted in Berns, *For Capital Punishment* (New York 1979), p. 22.
9. Ibid.
10. Sellin, *The Death Penalty*, (Philadelphia, 1959), p. 15.
11. Quoted in Berns, p. 48.
12. Gaylin, *The Killing of Bonnie Garland (New York, 1982),* p. 253.
13. Ibid., p. 341.

## 22. Capital Punishment: The Constitution and the Supreme Court

1. Berns, p. 31.
2. Ibid.
3. Bedau, p. 4.
4. Van Den Haag, "The Death Penalty Once More," *University of California-Davis Law Review* (Summer 1985) pp. 957-958.
5. *Furman v. Georgia,* 408 U.S. 238 (1972).
6. Ibid.
7. *Gregg v. Georgia,* 428 U.S. 153 (1976).
8. Ibid.
9. Ibid.
10. Berns, p. 189.

## 23. Society's Moral Outrage

1. Beccaria, p. 42.
2. Quoted in Berns, p. 139.
3. Dostoevsky, *Notes from Underground* (New York, 1969), p. 41.
4. *Gregg v. Georgia.*
5. Dougherty, *General Ethics*, (New York, 1959) p. 161.
6. Wilson and Herrnstein, *Crime and Human Nature*, (New York, 1985) p. 498.
7. von Husch, *Doing Justice: The Choice of Punishments*, (New York, 1976), p. 69.
8. Murphy, *Retribution, Justice and Therapy* (Holland, 1979) p. xi.
9. Bourke, *Ethics* (New York, 1966), p. 334.
10. Conat, *Justice at Nuremberg*, (New York, 1983), p. 106.
11. House and Yoder, *The Death Penalty Debate* (Dallas, 1991) p. 77.

# SELECT BIBLIOGRAPHY

Adams, Mark B., editor, *The Wellborn Science* (New York, 1990)

Adler, Mortimer, *The Common Sense of Politics* (New York, 1996)

Azar, Larry, *Man: Computer, Ape, or Angel?* (Boston, 1989)

Azar, Larry, *Twentieth Century In Crisis: Foundations of Totalitarianism* (Iowa, 1990)

Beccaria, Cesare, *On Crimes and Punishment* (Indianapolis, 1963)

Bedau, Henry Adam, *The Death Penalty in America* (New York, 1997)

Berns, Walter, *For Capital Punishment* (New York, 1979)

Bourke, Vernon, *Ethics* (New York, 1996)

Breitman, Richard, *The Architect Of Genocide* (New York, 1991)

Budziszewski, J., *True Tolerance* (New Jersey, 1992)

_____, *Written on the Heart: The Case For Natural Law* (Illinois, 1997)

*Catechism of the Catholic Church* (San Francisco, 1994)

*Code of Canon Law* (Canon Law Society of America, 1983)

Conot, Robert, *Justice at Nuremberg* (New York, 1983)

Conrad, John and Van Den Hagg, *The Death Penalty: A Debate* (New York, 1983)

Coyne, Randall and Entzeroth, Lyn, *Capital Punishment and the Judicial Process* (North Carolina, 1994)

Cromartie, Michael, editor, *A Preserving Grace* (Washington, DC, 1997)

Dostoevsky, *Notes from Underground* (New York, 1969)

Dougherty, Kenneth, *General Ethics* (New York, 1959)

Fisher, Klaus, *Nazi Germany: A New History* (New York, 1995)

Friedlander, Henry, *The Origins of Nazi Genocide* (Chapel Hill, 1995)

Friedlander, Saul, *Nazi Germany and the Jews* (New York, 1997)

Gallagher, Hugh Gregory, *By Trust Betrayed* (Vadamere, 1995)

Gaylin, Willard, *The Killing of Bonnie Garland* (New York, 1982)

Germino, Dante, *Machiavelli to Marx* (Chicago, 1972)

Gomez, C., *Regulating Death: Euthanasia and the Case of the Netherlands* (New York, 1991)

Grisez, Germain and Shaw, *Beyond the New Morality* (Indiana, 1988)

Gulan, Richard, *Euthanasia* (Paulist Press, 1994)

Hendrin, Herbert, *Seduced by Death* (New York, 1997)

_____, *Suicide in America* (Norton, 1995)

Hobbes, Thomas, *Leviathan* (London, 1841)

_____, *Philosophical Rudiments Concerning Government and Society* (London, 1841)

Hofstader, Richard, *Social Darwinism In American Thought* (Beacon Press, 1955)

House, Wayne and Yoder, John, *The Death Penalty Debate* (Dallas, 1991)

Johnson, Paul, *Modern Times* (New York, 1993)

Kevorkian, Jack, *Prescription Medicide: The Goodness of a Planned Death* (New York, 1991)

Kimbrell, Andrew, *The Human Body Shop* (San Francisco, 1993)

Kinter, Earl, editor, *The Haldamar Trial* (London, 1949)

Koosed, Margery, *Capital Punishment, Volume 1: The Philosophical, Moral and Penological Debate over Capital Punishment* (New York, 1996)

Kreeft, Peter, *Making Choices: Finding Black and White in a World of Grays* (Ann Arbor, 1990)

Kuhl, Stefan, *The Nazi Connection: Eugenics, American Racism* (Oxford, 1994)

Lifton, Robert Jay, *The Nazi Doctors* (Basic Books, 1986)

Maestri, William, *Choose Life and Not Death* (Alba House, 1985)

Maritain, Jacques, *Christianity and Democracy and the Rights of Man and the Natural Law* (San Francisco, 1986

_____, *The Person and the Common Good* (Indiana, 1966)

Marx, Paul, *And Now Euthanasia* (Human Life International, 1985)

McCarthy, Donald, *Moral Responsibility In Prolonging Life Decisions* (Pope John XXIII Center, 1981)

Messner, Johannes, *Social Ethics and the Natural Law in the Western World* (St. Louis, 1949)

Miller, Richard, *Nazi Justiz* (Praeger, 1995)

Miner, Brad, *The Concise Conservative Encyclopedia* (New York, 1996)

Muggeridge, Malcolm, *Vintage Muggeridge* (Michigan, 1985)

Novak, Michael, *Free Persons and the Common Good* (New York, 1989)

O'Donnell, Thomas, *Medicine and Christian Morality* (Alba House, 1976)

Osterle, John, *Ethics* (Prentice Hall, 1956)

Proctor, Robert, *Racial Hygiene* (Harvard, 1988)

Rice, Charles, *Beyond Abortion: The Theory and Practice of the Secular State* (Chicago, 1979)

_____, *50 Questions on the Natural Law* (San Francisco, 1993)

Rommen, Heinrich, *The Natural Law* (St. Louis, 1947)

Sabine, George, *A History of Political Theory* (New York, 1963)

Sellin, Thorsten, *The Death Penalty* (Philadelphia, 1959)

Senander, Mary, *The Living Will* (Leaflet Co., 1993)

Sereny, Gitta, *Albert Speer: His Battle With Truth* (Knopf, 1995)

Sheen, Fulton, *Liberty, Equality and Fraternity* (New York, 1938)

Sheldon, Garrett, *The Political Philosophy of Thomas Jefferson* (Baltimore, 1991)

Sigmund, Paul, *Natural Law In Political Thought* (Maryland, 1971)

Simon, Yves, *The Tradition of Natural Law* (New York, 1992)

Singer, Peter, *Rethinking Life and Death* (New York, 1995)

Smith, Wesley, *Forced Exit* (Times, 1997)

Strauss, Leo, *Natural Right and History* (Chicago, 1953)

_____, *The Political Philosophy Of Hobbes* (Chicago, 1963)

Tusa, Ann and John, *The Nuremberg Trial* (New York, 1983)

Von Hirsch, A., *Doing Justice: The Choice of Punishments* (New York, 1976)

Weintraub, Lloyd, *Natural Law and Justice* (Harvard, 1987)

Wertham, Frederic, *The German Euthanasia Program* (Hayes, 1980)

Wilson, James Q. and Herrnstein, Richard, *Crime and Human Nature* (Simon and Schuster, 1985)

Wilson, James Q., *The Moral Sense* (New York, 1993)

# INDEX

*This book was designed and set into type*
*by Mitchell S. Muncy*
*at Spence Publishing Company,*
*Dallas, Texas,*
*and printed and bound by Thomson-Shore, Inc.*
*Dexter, Michigan.*

❖

*The jacket design is by Gwen V. Purtill*

❖

*The text face is Garamond,*
*supplemented by Goudy Old Style titling.*

❖

*The paper is acid free and is of archival quality.*